ARITHMETIC
FOR THE PRACTICAL WORKER

4th Edition

MATHEMATICS LIBRARY FOR PRACTICAL WORKERS

Arithmetic *for the Practical Worker*
Algebra *for the Practical Worker*
Geometry *for the Practical Worker*
Trigonometry *for the Practical Worker*
Calculus *for the Practical Worker*

A GROUP OF BOOKS THAT MAKE EASY THE HOME STUDY OF THE WORKING PRINCIPLES OF MATHEMATICS

ARITHMETIC
FOR THE PRACTICAL WORKER

4th Edition

MATHEMATICS LIBRARY FOR PRACTICAL WORKERS

VNR VAN NOSTRAND REINHOLD COMPANY
NEW YORK CINCINNATI TORONTO LONDON MELBOURNE

Library of Congress Catalog Card Number: 81-11587
ISBN: 0-442-28275-3

Manufactured in the United States of America

Published by Van Nostrand Reinhold Company Inc.
135 West 50th Street, New York, N.Y. 10020

Van Nostrand Reinhold Limited
1410 Birchmount Road
Scarborough, Ontario MIP 2E7, Canada

Van Nostrand Reinhold
480 Latrobe Street
Melbourne, Victoria 3000, Australia

Van Nostrand Reinhold Company Limited
Molly Millars Lane
Wokingham, Berkshire, England

15 14 13 12 11 10 9 8 7 6 5 4 3

Library of Congress Cataloging in Publication Data

Thompson, James Edgar, 1892-
Arithmetic for the practical worker.

(Mathematics library for practical workers)
Includes index.
1. Arithmetic--1961- I. Title.
II. Series.
QA107.T46 1981 513'.13 81-11587
ISBN 0-442-28275-3 AACR2

PREFACE TO FOURTH EDITION

The popular demand for this book and for the other four in the MATHEMATICS LIBRARY FOR PRACTICAL WORKERS has been maintained through three editions and for over fifty years. The publisher hopes that this Fourth Edition will prove as valuable to a new generation of practical workers who must work with mathematics on a regular, practical, non-theoretical basis.

PREFACE TO THIRD EDITION

Continued use of this book over the years has demonstrated the effectiveness of its method of presentation. This method has, therefore, been retained in this third edition. The changes have been directed toward the modernization of the general statements and problems, and especially toward the extension of the basic method to the treatment of the developments in arithmetic that have become important as a result of more recent progress in business and technology. Prominent among these is binary arithmetic, the counting method used in so many digital computers, which is fully explained, and illustrated, in Chapter 21. Other changes have been made to adapt the treatment to present-day needs.

PREFACE TO FIRST AND SECOND EDITIONS

ARITHMETIC is the foundation on which the structure of mathematics is erected. A knowledge of its principles is an essential prerequisite to the study of other and higher branches of mathematics. In addition, one must be able to handle arithmetical problems with ease and assurance to gain success in any trade or business.

This book is designed for those who wish to gain such facility by home study without the aid of a teacher. The book can also be used to advantage by those who wish to refresh their knowledge of the subject with a review of its principles or to learn the special applications of fundamental arithmetical operations to trade and business. It is one of a series of books on mathematics for home study, each volume of which is designed to assist those who wish to improve their ability to handle mathematical operations.

Commencing with a brief review of the fundamental operations of addition, subtraction, multiplication and division, the book proceeds to a study of the general and special rules of calculation. While not a book on finance or trade mathematics as such, it includes enough special applications to give a working knowledge of their fundamentals and to develop enough of a background so that the student may continue his studies through the medium of specialized textbooks.

It is believed that the introduction of logarithms and the inclusion of the several tables will add to the usefulness and interest of the book. For similar reasons many special methods, measures, rules, etc., which are of interest only to specialists in certain lines of work have been omitted. The material in Part III is intended to be more complete and detailed than in many elementary texts, because of its interest and because it is the basis of industrial applications. In every case a stated rule or method is amplified by illustrations and by careful explanation. The author has tried to keep continually in mind the fact that the book must give a practical working knowledge of the subject without the necessity of explanation by a teacher. It is therefore written in a personal and informal style, and contains a large number of problems and examples illustrative of the rules and methods given.

In this new second edition many new problems have been added, and some of these bear on uses and applications of arithmetic which the Second World War has brought into prominence. In this edition a number of other changes have been made, some in response to the suggestions of many readers of the first edition, and some to meet the need to bring the applications up-to-date and to improve the explanations in some parts. No major changes in plan, scope or content have been made, however, as so many readers expressed satisfaction with the book in its first edition. All errors which have been discovered have been corrected, and it is hoped that the new edition will be free from such small errors as are almost unavoidable in a first edition. Many of the original errors were pointed out by readers, and the author and publisher take this opportunity to thank readers who have pointed out such errors, and to ask new readers to inform us of any errors which they may discover in the new edition.

In the light of experience with the first edition it is felt that the serious reader who has carefully studied the contents of this book will be equipped to handle arithmetical problems which normally arise in ordinary business and will also have a sufficient knowledge of the basic principles of the subject to enable him to pursue the study of algebra as presented in the succeeding volume of this series.

J. E. THOMPSON

Brooklyn, N. Y.
October, 1945

CONTENTS

PART 1

INTRODUCTION AND REVIEW

PART 2

PRINCIPLES AND METHODS OF ARITHMETIC

Chapter 10 RATIO AND PROPORTION

Chapter 11 SERIES AND PROGRESSIONS

PART 3

MEASUREMENTS

Chapter 12 SYSTEMS OF COMMON MEASURES

Chapter 13 CALCULATION WITH DENOMINATE NUMBERS

Chapter 14 TIME, TEMPERATURE AND ANGLE MEASURE

PART 4

SOME SPECIAL APPLICATIONS

Chapter 18 GRAPHS

TABLES

Part 1

INTRODUCTION AND REVIEW

Chapter 1

INTRODUCTION

1. Historical Introduction. Early man probably did his counting by the aid of his ten fingers, and we still "count on our fingers" and by fives and tens. It is not certain what the first number symbols or numerals were. They were probably fingers held up, groups of pebbles or sticks, notches on a stick, etc., one for each article to be counted. Quite early, however, groups of marks, |, | |, | | |, | | | |, etc., were used to represent numbers, and these are still used as tallies in counting, as |, | |, | | |, | | | |, ~~| | | |~~ for 1, 2, 3, 4, 5, etc.

The earliest *written* symbols of the Babylonians were wedge-shaped or *cuneiform* symbols. The vertical wedge (V) represented *one*, the horizontal wedge (<) *ten*, and the two together (V <) one *hundred*. Other numbers were formed by writing these three symbols in various combinations.

The Egyptians used *hieroglyphics*, pictures of objects or animals that in some way represented the idea of the number they wished to represent. Thus *one* was represented by a vertical staff (|), *ten* by a symbol like a horse shoe (Ω), one *hundred* by a crook or spiral (℮), one *million* by the picture of a man with hands outstretched in an attitude of astonishment, etc. To represent other numbers these symbols were placed beside each other in the proper combinations and added.

The Greeks used the letters of their alphabet for number symbols and combined them after the manner of the Babylonians to form other symbols. The Romans also used the letters of their alphabet for number symbols and these have come down to us as I, II, III, IIII (or IV), V, X, etc.

None of the ancients had a symbol for zero and in general they found arithmetical calculation very cumbersome and difficult.

The system now used by almost all peoples of the civilized world originated in India with the Hindus. It was obtained from the Hindus by the Arabs, who introduced it into Europe soon after their conquest of Spain in the eighth century A.D. On this account the numerals or figures of this system are now called the *Arabic* numerals.

The numeral 1 is simply the single mark |. The 2 is formed from two joined horizontal marks Z and the 3 from three such marks ≡|. The 4, 5, 6, 7, 9 and probably 8 were the initial letters of the corresponding words for numerals in a language used in the north of India a few centuries B.C. The Arabs used the symbol 0 to mean *empty*. Their word for "empty" was "sifra" and from this we get our word *cipher* for the zero symbol.

The ancient peoples had no convenient signs of operation for numbers. Addition was generally indicated by placing together the numbers to be added, and other operations were stated in words. The signs + and − seem to have been used first by Widman, in a book on arithmetic published in Germany in 1489. He used them to mark excess or deficiency, as "more" or "less," but they soon came into use as signs of operation indicating "add" and "subtract." The sign × was used by William Oughtred in England in 1631 to indicate multiplication. The Arabs quite early indicated division in the form of fractions. Following this method Rahn used the sign ÷ to indicate division, in a book on the subject of Algebra which was published in 1669, the dot above the line indicating the position of the dividend or numerator in a fraction and the dot below the line that of the divisor or denominator. Robert Record in England introduced the equality sign (=) in 1557, two lines of equal length indicating equality. These signs are now used everywhere in all branches of mathematics.

2. Signs of Operation Used in Arithmetic. The signs most used in arithmetic to indicate operations with numbers are the *plus* (+), *minus* (−), *multiplication* (×), and *division* (÷) signs. When either of these is placed between any two numbers it indicates respectively that the sum, difference, product, or quotient of the two numbers is to be found. The *equality* sign (=) shows that any indicated operation or combination of numbers written before it (on the left) produces the result or number written after it. Thus

$$12 + 4 = 16, \quad 12 - 4 = 8, \quad 12 \times 4 = 48, \quad 12 \div 4 = 3.$$

The *fraction line*, a horizontal line with numbers above and below it, is also used to indicate division. Thus $\frac{12}{4}$, twelve fourths, is the same as $12 \div 4$ and we can write $\frac{12}{4} = 12 \div 4 = 3$. Similarly $\frac{5 \times 4}{2 \times 2} = \frac{20}{4} = 5$, which means that the product 5×4 is divided by the product 2×2.

Another sign sometimes used is the *parentheses* ( ). If any number or combination of numbers is inclosed in parentheses the whole combination, or result of the indicated operations, inside the parentheses is to be considered as a single number. Thus $3 \times (4 + 2)$ means 3×6, the $4 + 2$ inside the parentheses being equal to 6. Similarly $8 - (3 + 2) = 8 - 5 = 3$; $7 + (4 \times 8) = 7 + 32 = 39$; $\frac{(5 \times 4)}{(2 \times 2)} = \frac{20}{4} = 5$; etc.

Other signs are used for certain purposes; these will be introduced as needed. Those given here are the ones most used in elementary arithmetic.

In the use of the arithmetical signs of operation in succession, when parentheses are not used, the signs $\times$ and $\div$ must be used first, and the signs $+$ and $-$ then used. Thus in the following expression which involves combinations of the several signs: $42 - 6 \div 3 + 4 \times 2 - 18$, the signs $\times$ and $\div$ give $4 \times 2 = 8$ and $6 \div 3 = 2$, and the expression becomes $42 - 2 + 8 - 18$. From this the final result is found by subtracting 2 from 42, then adding 8 and finally subtracting 18, giving 30. Similarly $5 + 2 \times 3 - 15 \div 5 + 4 = 5 + 6 - 3 + 4 = 12$, the results of 2×3 and $15 \div 5$ being found before they are added or subtracted.

3. Writing Very Large and Very Small Numbers. A very large number such as 1,000,000,000, which is 1 followed by 9 zeros is frequently written in the form 10^9, the small 9 above and to the right of the 10 indicating the number of zeros after the 1. Similarly 2,500,000 which is $2\frac{1}{2}$ millions or $2.5 \times 1,000,000$, would be written 2.5×10^6; 38970000000 as 3.897×10^{10}; etc.

A very small number, expressed as a decimal fraction, such as .0000007, may be written as $.7 \times 10^{-6}$ or 7×10^{-7}. The -6 and -7 indicate the number of decimal places that the decimal point has been moved to the right. Similarly, .00000000425 may be written as $.425 \times 10^{-8}$, or preferably as 4.25×10^{-9}. Also, .00726 is 7.26×10^{-3}.

The method of writing very large or very small numbers by means of 10 with a plus or minus number, respectively, in small figures just above and to the right of the 10, is referred to as the method of *powers of* 10 and the number in small figures is called the exponent of 10, for reasons which we shall see later. This method is much used in scientific calculations.

4. Use of Letters for Numbers. The initial letters of words are sometimes used to represent the number for which the word stands. Thus if the *price* (p) per pound of meat is .80 dollars, and 4 pounds *weight* (w) is purchased, the total *cost* (c) is found by multiplying the price by the weight, $3.20 = 4 × .80. In general *cost* = (*weight*) × (*price*), or in symbols $c = w \times p$. In the same manner we can indicate that the *area* (A) of a rectangle is equal to the product of its *length* (L) and *width* (W) by writing *area* = (*length*) × (*width*) or in symbols $A = L \times W$. If N represents one number and n another, the quotient of the first by the second is $Q = \frac{N}{n} = N \div n$; the sum of any two numbers being equal to a third may be written $a + b = c$; etc.

This method of using letters as symbols for numbers is very useful and much used in the writing of certain rules, and we shall occasionally find it very convenient. The method is used very largely in algebra and in the application of mathematics to scientific and engineering theories and calculations.

5. Concrete and Abstract Numbers. If a number is applied to specific counted or measured *things*, *objects*, or *quantities*, as 7 *books*, 5 *pencils*, a length of 4½ *feet*, a weight of 175 *pounds*, a value of 5.25 *dollars*, etc., the number is said to be a *concrete* or sometimes a *denominate* number.

When numbers do not so apply to things they are called *abstract* numbers. Thus numbers at random used in counting, adding, subtracting, etc., as 1, 2, 3, 4, 5 or 4275 + 982 = 5257, ½ × 4496 = 2248, etc., are abstract numbers.

Abstract numbers, numbers used simply *as* numbers, are studied and used in all the rules of pure arithmetic. Concrete numbers used in counting and measuring things, dimensions, values, etc., are used in the applications of arithmetic to solve problems, in the calculations of shop work, business, science, etc. Thus the simple multiplication 4 × .80 = 3.20 involves only abstract numbers, but the example (4 lbs.) × (80¢ per lb.) = $3.20 involves concrete numbers.

Chapter 2

THE FUNDAMENTAL OPERATIONS

6. Addition. It will be recalled by the reader that the process of finding the combined amount of two or more numbers is called *addition* and this combined amount is called the *sum* of the several numbers. To find this sum we are said to *add* the numbers.

In order to add several numbers, such as 4567, 832, 98122, 65, for example, the numbers are written one beneath the other with their figures or *digits* in columns and the last or right-hand digits in the *same* column, thus:

```
 4567
  832
98122
   65
```

The figures in each column are then added, beginning at the right. If the sum of the figures in any one column is a number of more than one figure, the last or right-hand figure of this number is written below the same column and the remaining figure or figures considered as another number and added to the next column to the left. This column is added as was the first, and the process is repeated and continued until all the columns have been added. The complete sum of the last column at the left is written down and the complete resulting number so formed of the column sums is the sum of all the numbers.

Applying this process to the four numbers given above we have for the sum of the first column $5 + 2 + 2 + 7 = 16$. Writing down the 6 and adding the 1 to the next column, the next sum is $1 + 6 + 2 + 3 + 6 = 18$. Writing the 8 at the left of the 6 already written down, and adding the 1 to the next column, the next sum is $1 + 1 + 8 + 5 = 15$. Writing the 5 at the left of the previous 86, giving 586, and adding the 1 in the next column, the next column sum is $1 + 8 + 4 = 13$. Writing the 3 as before, giving 3586, and adding the 1 in the next column, the last column sum is $1 + 9 = 10$. Writing this down in full, the complete sum is 103,586. The complete operation is written out in full as follows:

4567
832
98122
65
103586

In order to keep track of the separate column totals and as an aid in checking the results, it is convenient to write the separate column sums at one side as well as writing the last figure of each under its own column. This column sum is written first; the last figure is then written under its column and the others carried over in each case to the next column. This procedure is illustrated in the following example.

To add:	47896	1st column:	40
	83924	2nd "	37
	1785	3rd "	46
	46751	4th "	27
	928	5th "	18
	54		
	9247		
	21425		
Sum =	212010		

To check the work, the columns should all be added again, beginning at the top and comparing the column sums with those set down at the right. As a separate check, the column sums may also be written in the following order and added as shown:

(1)	40
(2)	37
(3)	46
(4)	27
(5)	18
Sum =	212010

Bookkeepers, accountants and scientific computers use many other short cuts and checks but these will not be discussed here.

7. Subtraction. Subtraction is the process of finding the difference between two numbers by taking the smaller of the two from the larger. The number which is subtracted is called the *subtrahend*, the other number is called the *minuend*, and their difference, the number which remains after the subtrahend is subtracted from the minuend, is called the *remainder*.

In order to find the remainder, the subtrahend is written under the minuend as in addition. Beginning at the right each figure in the subtrahend is subtracted from the figure above it and the individual remainder written below in the same column. When the process is completed the third complete number, below the subtrahend, is the desired remainder.

As an example, it is required to subtract 42356 from 98577. The separate operations are $7 - 6 = 1$, $7 - 5 = 2$, $5 - 3 = 2$, $8 - 2 = 6$, $9 - 4 = 5$, giving 56221 as the remainder. The operation is written out in full as follows:

Minuend: 98577
Subtrahend: 42356
Remainder: 56221

As a check the remainder and the subtrahend should be added. If the subtraction was correct the sum of this addition will be equal to the minuend.

If any figure in the subtrahend is a number greater than the one above it in the minuend, it cannot be subtracted directly and the following method is used. A single unit (1) is "borrowed" from the next figure to the left in the minuend and written (or imagined to be written) before the figure which is too small. The figure of the subtrahend is then subtracted from the number so formed and the remainder figure written down in the usual way.

The minuend figure from which the 1 was borrowed is now considered as a new figure, 1 less than the original, and its corresponding subtrahend figure subtracted in the usual way. If the minuend figure is again too small, the process just described is repeated.

As an illustration of the procedure just described, let it be required to subtract 26543 from 49825. The operation is written out as follows:

7_1
Minuend: $49\not{8}25$
Subtrahend: 26543
Remainder: 23282

Here the subtrahend figure 4 is subtracted from 12 instead of the original 2, and the subtrahend figure 5 is then subtracted from 7 instead of the original 8. One more example is given:

$5_1 4_1$
Minuend: $987\not{6}\not{5}4$
Subtrahend: 234567
Remainder: 753087

In such cases as these, the result is also to be checked by adding the remainder and the subtrahend.

8. Multiplication. If a certain number is written down four times, one below the other, and the four equal numbers are added, the sum is said to be four *times* that number. The original number is said to be *multiplied by* 4 and the result is called the *product* of that number and 4. In the same way, if any given number is counted any specified number of times the first number is said to be multiplied by the second and the result is the product of the two numbers. The number which is multiplied is called the *multiplicand* and the number by which it is multiplied is called the *multiplier.*

Table I at the end of this book shows the products of every two numbers from 1 to 20. This is one form of the ordinary "multiplication table," extending to 20 instead of the usual 12. The product of any number in the first column by any number in the first line is found where the line and column meet, which contain the two numbers. The complete table for the 3's, the 9's, the 17's, etc., is given in either the line or the column beginning with 3, 9, 17, etc.

From the multiplication table, $3 \times 9 = 27$. This is read "3 times 9 is 27"; 9 is the multiplicand, 3 the multiplier, and 27 the product. Similarly, $4 \times 12 = 48$, and this suggests the method of handling any multiplicand of several figures. Thus, it is noted that in the product $4 \times 12 = 48$, the 8 is 4×2 and the 4 is 4×1. Using this method, $3 \times 23 = 69$, and similarly, $4 \times 521 = 2084$, which product is obtained by beginning at the right and multiplying $4 \times 1 = 4$, $4 \times 2 = 8$, $4 \times 5 = 20$, and writing down the separate results as in the addition of columns.

If any partial product consists of two figures, the right-hand figure is written down and the other carried over and added to the next separate product as in adding columns. For example, to multiply 4137 by 5: Beginning at the right, $5 \times 7 = 35$; put down the 5 and carry over the 3 to the next product, $5 \times 3 = 15$, giving 18 when added. Write the 8 before the previously written 5, giving 85, carry over the 1 to the next product, $5 \times 1 = 5$, giving 6, and write this 6 before the previous 85, giving 685. The final product is $5 \times 4 = 20$, and when this is written before the 685 the complete product 20685 is obtained. Therefore, $5 \times 4137 = 20{,}685$. The complete operation is written out as follows, the multiplier being written under the last figure of the multiplicand, the last figure of each separate product being written under the corresponding figure of the multiplicand, and each separate product being written at the side as in addition, for a check if desired.

Multiplicand:	4137	(1)	35
Multiplier:	5	(2)	15
Product:	20685	(3)	5
		(4)	20
		Sum =	20685

If the multiplier contains more than one figure the multiplicand is multiplied by each of those figures in turn, beginning at the right, and each multiplication being carried out as just described for the one-figure multiplier. The successive products are written one below the other, each beginning one space farther to the left than the preceding, and the several products are then added.

As an example, the number 4137, just multiplied by 5, will be multiplied by 745, the steps being indicated.

```
                 4137
                  745
Multiply by 5,   20685
  "      " 4,   16548
  "      " 7,  28959
          Add, 3082065 = final product.
```

In the same manner a number consisting of any number of figures may be multiplied by any other number containing any number of figures. In finding the product of any two numbers which do not contain the same number of figures, the one with the lesser number of figures should be taken as the multiplier.

9. Division. The process of finding how many times one number is contained in a second number is called *division*. The first number is called the *divisor*, the second is called the *dividend* and the result is the *quotient*. Thus 12 is contained in 72, 6 times. In getting this result, 72 is said to be *divided by* 12 and the operation is indicated by writing $72 \div 12 = 6$. Here, 72 is the dividend, 12 the divisor and 6 the quotient.

Division is the operation inverse to multiplication and corresponding to the common multiplication table there is also a "division table."

$$7 \times 5 = 35$$
$$35 \div 5 = 7$$

Division and multiplication are called inverse operations because division "undoes" what multiplication "does." Table I at the end of this book is a division table. The quotient of any number in the body of the table divided by the number on the same line in the first column is given at the top of the column containing the dividend.

For dividends and divisors consisting of several figures use is made of such a division table and the complete division is carried out in steps or stages corresponding to the successive steps in multiplication. According as the divisor is less than 21 or greater than 20 (when Table I is used) one or the other of two general methods is used. These two methods are known as *short division* and *long division.*

In short division the divisor is divided directly into each successive figure or pair of figures of the dividend, beginning at the left, and the successive separate quotients written down one after the other to form a number which is the complete quotient.

As an illustration, let us divide 129636 by 3. As 3 is not contained in 1 a whole number of times the first two figures, 12, are taken as a partial dividend. Then $12 \div 3 = 4$, $9 \div 3 = 3$, $6 \div 3 = 2$, $3 \div 3 = 1$, and the complete quotient is 43212. That is, $129636 \div 3 = 43212$. This operation is usually written out in the following form: $\dfrac{3)129636}{43212}$ and the separate quotients are written below the corresponding separate dividends.

If any of the separate dividends (figures of the complete dividend) are smaller than the divisor, or do not contain the divisor exactly a whole number of times, a slightly different procedure is followed, as illustrated by the following example: Divide 189576 by 9. We get $18 \div 9 = 2$ and $9 \div 9 = 1$, but 9 is not contained in 5, so a zero is written for the next figure of the quotient. The next individual dividend is then taken as 57 instead of 7, and $57 \div 9 = 6$ with a remainder of 3 (since $6 \times 9 = 54$). This 3 is written (or imagined written) before the next figure, 6, giving 36, and this 36 is used as the next dividend. The final figure of the quotient is then $36 \div 9 = 4$, the successive figures of the complete quotient are 2, 1, 0, 6 and 4. The result is, therefore, $189576 \div 9 = 21064$. The complete operation is written as follows: $\dfrac{9)189576}{21064}$.

The division 3082065 ÷ 745 is carried out by long division according to the following scheme, the explanation following the work:

```
745)3082065(4137
    2980
    ----
     1020
      745
     ----
      2756
      2235
      ----
       5215
       5215
       ----
```

The first figure of the divisor is mentally and tentatively divided into the first figures of the dividend, 30, and the quotient, 4, is written at the right as the first figure of the desired final quotient. The *entire* divisor, 745, is then multiplied by this first figure 4 and the product 2980 written below the first four figures 3082 of the dividend. The 2980 is then subtracted and the remainder 102 written in the usual manner for subtraction. The next figure of the dividend, 0, is written after this 102 to form a new partial dividend, and the 7 of the divisor mentally divided into the first figures, 10, of this dividend. The quotient 1 is written beside the 4 already obtained as the second figure of the quotient, and the divisor 745 multiplied by it. The product 745 is written under the partial dividend 1020 and subtracted as before. The remainder is 275 and the next figure 6 of the dividend is copied down after it to give a new partial dividend, 2756. The 7 of the divisor is divided into the 27 of this partial dividend, giving a partial quotient 3 as the next figure of the complete quotient, and the preceding steps are repeated. This process is continued until there is no remainder in the last subtraction and no more figures in the dividend to write down. The division is then complete and the complete quotient is 4137. Therefore, 3082065 ÷ 745 = 4137.

Since division and multiplication are inverse operations, division can always be checked by multiplying the quotient by the divisor. If there are no errors the product obtained will be the original dividend. In the example just worked out the quotient 4137 has already been multiplied by the divisor 745 in article 8 and the product was 3082065, the dividend in the present example. The division just carried out is, therefore, correct.

If in any case in long division the figure of the quotient obtained by dividing the first figure of the divisor into the first figures of the partial dividend, gives on multiplication by the divisor a product larger than the partial dividend, the next smaller figure or number must be taken as the partial quotient.

If in any case, after writing the next figure of the dividend beside the previous remainder to form the next partial dividend, this partial dividend is less than the divisor, a zero (0) is written as the next figure of the quotient, and another figure of the dividend written beside the last one copied down, to form a new partial dividend for the next partial division.

If in any case, in long or short division, there is a final remainder when the last figure of the dividend has been used, this remainder is written as a numerator over the divisor as a denominator to form a fraction, and this fraction, or the fraction reduced to its lowest terms, is written after the last figure of the quotient already obtained as a final part of the complete quotient.

The last several cases are illustrated in the following example:

$$258)104874392(406489\tfrac{115}{129}$$

```
258)104874392(406489 115/129
    1032
    ----
      1674
      1548
      ----
       1263
       1032
       ----
        2319
        2064
        ----
         2552
         2322
         ----
          230
```

$$\frac{230}{258} = \frac{115}{129}$$

After the first subtraction, the 7 is drawn down from the dividend to form the partial dividend 167 but this is less than the divisor 258. The corresponding figure of the quotient is therefore 0, and the next figure of the dividend is drawn down, 4. When all the figures of the dividend have been used the last remainder is 230. Using this as numerator and 258 as denominator, the fraction $\frac{230}{258}$ is formed, which when reduced to its lowest terms becomes $\frac{115}{129}$. This is added to the quotient already obtained to give the complete quotient. The result is, therefore,

$$104874392 \div 258 = 406489\tfrac{115}{129}.$$

Chapter 3

CALCULATION WITH DECIMALS

10. Decimal Numbers. We recall here a few definitions and properties of fractions which are no doubt familiar to the reader and which are essential to the study of decimals. They will be treated in more detail in Chapter 6 when we come to study the properties and handling of fractions fully.

A *fraction* is a part of a quantity which is treated as a whole or a unit. In arithmetic, a proper fraction is a number which represents a part, that is, a number which is less than 1. In writing a common fraction, two numbers are used, called the *numerator* and *denominator.* The denominator names the fractional part and the numerator indicates how many of those parts are used. Thus, two of the three equal parts called "thirds" are indicated by the fraction $\frac{2}{3}$; similarly, seven ninths are indicated by $\frac{7}{9}$; etc. The number above the line is the numerator and that below is the denominator.

If the denominators are tenths, hundredths, thousandths, etc., the numerators may be written without the denominators and with a period placed before the numerators. In the fraction so written the numerator of the tenths is placed first after the period, that of the hundredths is second, thousandths next, etc. Thus, .7 is $\frac{7}{10}$ (seven tenths); .06 is $\frac{6}{100}$ (six one hundredths); .23 is $\frac{23}{100}$; .009 is $\frac{9}{1000}$; .475 is $\frac{475}{1000}$; .0785 is $\frac{785}{10000}$; etc. Fractions written in this manner are called *decimal* fractions, and the period placed before the figures is called the *decimal point.*

If any number consists of a whole number and a fraction, as $18\frac{7}{10}$, the whole number is written before the decimal point and fraction after, thus: 18.7. Similarly, $293\frac{69}{100} = 293.69$; $12\frac{63}{1000} = 12.063$; etc. Such numbers are called *decimal numbers* or simply *decimals.*

The decimal system is the basis of the money system of the United States and of most other countries. Facility in handling decimals is important for that reason as well as because of their mathematical significance.

In calculations, decimal numbers are used like whole numbers, the only difference being that the position of the decimal point must be carefully observed and attended to.

Assuming that the reader has a little knowledge of decimals, a brief review of the methods of calculation with decimals is given in this chapter. A full explanation of the conversion of common fractions into decimals is given later in the chapter on FRACTIONS.

11. Addition and Subtraction of Decimals. In adding and subtracting decimals, the numbers are written one below the other *with their decimal points in line* and the addition or subtraction is then carried out in the usual way, the same as for other numbers. In the sum or remainder, the decimal point is placed in line with the decimal points above.

As an illustration, let us add 1254.62, 129.36, 26375.8, 4.753, 18.9654. This is written down and carried out at the right: The sum is "27,783 and 4,984 ten thousandths."

```
 1254.62
  129.36
26375.8
    4.753
   18.9654
----------
27783.4984
```

As another illustration, we show the addition of the decimal fractions .7, .08, .753, .75, .003, .045, .12345. This addition is carried out below at the left and the sum is "2 and 45, 445 hundred thousandths."

```
 .7
 .08
 .753
 .75
 .003
 .045
 .12345
-------
2.45445
```

The subtraction 42.963 − 18.356 is performed as follows:

```
42.963
18.356
------
24.607
```

and the remainder is "24 and 607 thousandths."

If the decimal fractional parts of the minuend and subtrahend do not have the same number of decimal *places* to the right of the decimal points, the blank places are considered as zeros. As an example, the subtraction 976.5928 − 854.937 is performed as at the left below; the subtraction 12.283 − 10.2617 is carried out at the right:

```
976.5928        12.283
854.937         10.2617
--------        -------
121.6558         2.0213
```

From these illustrations it is clear that addition and subtraction of decimals will give no trouble when the numbers are written *with decimal points in line, one below the other.*

12. Multiplication of Decimals. In multiplying two decimal numbers, they are written down without regard to the position of either of their decimal points and the multiplication is carried out *exactly* as in the case of ordinary whole numbers. The position of the decimal point in the product is then determined by the

RULE: *The number of figures to the right of the decimal point in the product of two decimal numbers equals the sum of the numbers of figures to the right of the decimal points in multiplicand and multiplier.*

The use of this rule is illustrated in the following multiplication: 263.42 × 12.4 = 3266.408. There are *two* figures to the right of the decimal point in the multiplicand (.42) and *one* in the multiplier (.4). In the product, therefore, according to the rule, there are 2 + 1 = 3 figures and the decimal point is placed before the 4.

```
   263.42
     12.4
  -------
   105368
   52684
  26342
  -------
  3266.408
```

In every case the number with the lesser number of figures altogether is used as the multiplier.

```
   .1468
    .463
 --------
    4404
   8808
  5872
 --------
.0679684
```

As a second example, let us multiply the two decimal fractions: .1468 × .463. This is done at the left. Here there are *four* figures to the right of the decimal point in the multiplicand (.1468) and *three* in the multiplier (.463). In the product, therefore, there must be 4 + 3 = 7 figures to the right of the decimal point. But the product contains only 6 figures as found in the usual way (679684). A zero is, therefore, added *before* these six figures to make up the required seven figures to the right of the decimal point, and the decimal point is placed before the zero. The product is, therefore, .0679684. That is,

$$.1468 \times .463 = .0679684.$$

The rule of decimal points in multiplication applies without exception.

13. Division of Decimals. Division of one decimal number by another is carried out without regard to the position of either of their decimal points in *exactly* the same manner as for whole numbers. The position of the decimal point in the quotient is then determined by the following.

RULE: *Subtract the number of places to the right of the decimal point in the divisor from the number in the dividend. The remainder is the number of figures to the right of the decimal point in the quotient.*

The application of this rule is illustrated in the following division: 679.684 ÷ 14.68 = 46.3. There are *three* decimal places in the dividend (.684) and *two* in the divisor (.68). The number of decimal places in the quotient is, therefore, 3 − 2 = 1 and the decimal point is placed before the last figure.

```
14.68)679.684(46.3
      587 2
      -----
       92 48
       88 08
       -----
        4 404
        4 404
        -----
```

As in multiplication, this rule for division applies without exception to either decimal fractions or mixed decimal numbers.

If a division involving decimals does not give an exact quotient, the remainder is not taken as the numerator of a fraction to form a part of the quotient as in the division of whole numbers, but a zero is added to the remainder to form a new partial dividend and the division is continued. This process is repeated until the quotient is exact or until a zero is obtained as a figure in the quotient, or as long as desired if the quotient does not come out exact. In this way the remainder fraction is automatically converted into a decimal fraction and included as a part of the remainder. Each added zero is counted as a decimal place in the dividend, and the rule applied as usual. This procedure is illustrated by the following example: 8293.7 ÷ 32.74 = 253.32.

```
32.74)8293.7    (253.320
      6548
      ------
      1745 7
      1637 0
      ------
       108 70
        98 22
       ------
        10 480
         9 822
        ------
           6580
           6548
           ----
            320
```

The last remainder is 32 and when a zero is added the partial dividend is 320. But 3274 will not go into 320 a whole number of times, and the next figure in the quotient is a zero. The division is, therefore, stopped and the rule applied to locate the decimal point in the quotient. After the last figure 7 was used in the dividend four zeros were used as additional figures. These four together with the original decimal figure (.7) make *five* decimal places (.70000) while there are *two* in the divisor (.74). There are, therefore, $5 - 2 = 3$ decimal places in the quotient, and the decimal point is placed before the second 3, giving as quotient 253.320. Since $.320 = \frac{320}{1000} = \frac{32}{100} = .32$, it is not necessary to retain the final zero and the quotient can be written 253.32.

Chapter 4

APPROXIMATE RESULTS IN CALCULATION

14. Approximate Results. In technical measurements, the precision or degree of refinement of the readings is limited by the nature of the measurement and the type of measuring instrument or scale that is used, and is never absolutely exact. For example, in readings on a rule divided into sixteenths of an inch, any length less than a sixteenth must be estimated, and if a mark on a line which is being measured falls between lines on the scale, its position is only estimated and is not known exactly. Similarly, in using a fine micrometer it is usually said that a measurement is made to "within a thousandth" or "within one ten thousandth" of an inch, but no closer, that is, to the third or fourth decimal place, but no closer.

In certain calculations, also, as in calculating the square root of 2, say, the result (as we shall see later) is not exact but can be carried to as many decimal places as desired. Approximately the result is 1.4142+, the plus sign indicating that other figures are to be added. Similarly, the circumference of a circle is (as we shall see later) 3.14159+ times the diameter, but this number cannot be calculated exactly. In both these cases the numbers can be calculated to any desired degree of approximation, but cannot be expressed *exactly*. Again, some common fractions cannot be expressed exactly as decimal fractions, but may be expressed as nearly exactly as desired.

Thus, many numbers, whether found by measurement or calculation, are necessarily inexact and in some cases cannot be carried beyond a certain degree of exactness. It is, therefore, often unnecessary, in fact undesirable, to carry calculations beyond a certain point. In particular, if the original numbers on which a calculation is based are not expressed to a precision beyond a certain point, any results calculated from these numbers to a greater precision have no meaning and such a calculation is a waste of time and may be misleading.

The question naturally arises as to how the precision or relative significance of the figures in a measurement or a calculation is to be determined or expressed. This question is discussed in the next article.

15. Significant Figures. In the number 3271 the figure 1 expresses units; the figure 7 expresses tens; the 2 expresses hundreds; and the 3, thousands. The number can, therefore, be written 3000 + 200 + 70 + 1. The 1 is, therefore, $\frac{1}{3271}$ of the number; each unit in the place of the 7, or third place, represents $\frac{10}{3271}$ of the number; each unit in the second place, $\frac{100}{3271}$; and each unit in the first place, $\frac{1000}{3271}$ of the number.

If, therefore, 3271 is the true value of a certain measured or calculated result, and 3270 the result as measured or calculated, the error is one part in 3271 parts, or $\frac{1}{3271}$ of the true value. In this case the result is said to be *correct to three figures*, the three figures 327 being known to be correct and the fourth figure uncertain. The fourth figure simply indicates that the full value of the approximate result is to be 3270 instead of 327, but as the first three are the only ones *known* to be correct, they alone are significant in stating the *relative* precision of the result.

If the number is 32.71 instead of 3271, the 1 now represents 1 hundredth instead of one unit. The 1 is still $\frac{1}{3271}$ of the total, however, and, as before, the 7 represents $\frac{70}{3271}$ of the total, etc. If, in this case, the result is expressed approximately as 32.7 it is still given to three *significant figures.*

Similarly, if 3271 is expressed as 3200 or 32.71 as 32, there are two significant figures and the relative error is 71 parts in 3271; and the same is true for 3.271 expressed as 3.2. To express either of these more correctly to two significant figures, however, it should be written as 3300 for 3271, 33 for 32.71, or 3.3 for 3.271. The same holds true for the decimal point in any position and the relative accuracy of a measured or calculated result depends on the *number of figures*, other than zeros at the beginning or end of the number (as in .003271 or 3271000), and *not* on the number of decimal places. Thus, a number such as 3271.31276315 may represent a fictitious precision which is impossible in actual measurement, and the result would in such a case probably be better expressed as 3271 which is correct to four figures.

It is to be emphasized and must be remembered that "approximate" does not necessarily mean "only partially correct" but is generally applied to mean "as accurate as desired or possible" and may indicate in certain cases extreme precision or accuracy. Thus, the number represented by 3.14159+ is approximately expressed by these six figures and though it has been calculated to over seven hundred figures, twelve decimal places are sufficient to calculate the circumference of the sun to within one tenth of an inch. (The diameter of the sun is nearly a million miles.)

16. The Last Figure of a Decimal. It is sometimes required to determine the value of a certain decimal fraction *correct to a certain definite number of decimal places.* Thus, by the method of article 34, $\frac{23}{24} = .95833+$. Expressed to *four* decimal places the real value of this fraction lies between .9583 and .9584; .9583 is .00003+ less than the true value, and .9584 is .00006+ greater. Therefore, .9583 is nearer the correct value and is said to be *correct to four decimal places.* Similarly, .958 is correct to three places and .96 to two.

In writing .95833+ as .9583 the second 3 is disregarded, or rejected; it is less than 5. In writing .9583 as .958 the last figure, 3, is rejected; it is *less than* 5. In writing .958 as .96 however, the rejected figure (8) is *greater than* 5 and the last figure retained (5) is *increased by* 1 *unit.* This is because .960 is nearer .958 than is .950. In general, we have the

RULE: *If a rejected figure in a decimal fraction is less than* 5 *it is disregarded; if it is* 5 *or greater the next figure to the left of it, if retained, is increased by* 1 *unit.*

Thus, the decimal .23649 correct to four places is .2365. To three places, however, we would not write this as .237 but as .236, because in .23649 the next figure after the 6 is 4, which is less than 5. Similarly, 3.1415926+ to six significant figures or five decimal places is written 3.14159; to five figures or four decimals it is 3.1416, which is sufficiently accurate for most purposes.

17. Approximation in Addition and Subtraction. As an illustration of approximation in addition, let us add the decimal fractions .234673, .322135, .114342, .563217, each being known to be *correct to six figures.* The addition gives 1.234367, but it is not certain that this result is correct because, in accordance with the discussion of the preceding article, each of the given decimals may be either greater or less than the true value by as much as .0000005 (due to rejection of the seventh figure). Since there are four numbers the total error in the sum may be as much as $4 \times .0000005$, or .000002. That is, the true sum may be as much as 1.234369 or as little as 1.234365. In either case the result *correct to six figures* would be written 1.23437.

In some cases, the next-to-last figure may be incorrect as shown in the following example: Add .131242, .276171, .113225, .342247, each fraction being correct to six figures. The apparent sum, by direct addition, is .862885 and as in the preceding example the true sum may be anything between .862883 and .862887. If it were known to be .862886 or .862887 the sum correct to five figures would be written .86289, but if the true sum were known to be .862883 or .862884 it would be written .86288, to five figures. It is not known, therefore, whether the apparent sum .862885 should be written, to five figures, as .86288 or as .86289, and thus the next-to-last figure in the apparent sum is not known to be correct. The result *correct to four figures*, however, is .8629 in either case.

A similar analysis of the sum of 5.866314, 3.715918, 0.568286, 4.342233, each number being correct to seven figures, will show that the third from last figure in the sum is uncertain. The result can still be expressed correct to five figures, however, and is 14.493, which the reader should verify.

As another illustration, let us subtract .238647 from .329528, each fraction being correct to six places. The apparent remainder is .090881. As each of the two given numbers may be in error by .0000005, but no more, their difference can *not* be in error by more than $2 \times .0000005 = .000001$. Hence, the remainder must be between .090880 and .090882 and the result correct to five places is .09088.

Cases may arise in subtraction, however, in which the second or third figure from the last is in doubt.

18. Illustrations of Approximation in Multiplication and Division. From the examples in addition and subtraction given in the preceding article it will be noticed that in many cases it is necessary to carry the result to two figures more than are required in order to determine exactly the last required figure. This will also, of course, be true in multiplication and division, because multiplication is simply extended addition, and division is the inverse of multiplication. As an example, let us find the product 3.14159×3.14159 correct to four decimal places. By multiplying in the usual manner the apparent result is found to be 9.8695877281. This product may also be found in the reverse order of that usually followed, that is, by beginning at the *left* and writing each partial product one place farther to the *right,* as shown below.

```
3.14159
3.14159
-------
9.42477
  .314159
  .1256636
  .00314159
  .001570795
  .0002827431
------------
9.8695877281
```

The result correct to four places is, as in article 16, 9.8696. It is, therefore, only necessary to know two figures more than these and it is, therefore, not necessary to write down the last four figures in the product, 7281. Hence it is not necessary to write the corresponding figures in each of the partial products and the multiplication can be carried out in the following contracted form (omitting also the decimal points and zeros in the partial products), writing each partial product correct to six decimal places.

```
          3.14159
          3.14159
          -------
          9.42477
            314159
            125664
              3142
              1571
               283
          --------
Product = 9.869589
        = 9.8696     to 4 decimal places.
```

The same procedure can be followed in any multiplication in which it is not necessary to write all the obtainable figures in the product.

In order to obtain a result correct to a specified number of figures in division, it is only necessary to carry the quotient to two more places than are required and then reject the last two in the usual way.

This process can be shortened in some cases, as illustrated in the following example: Divide 9376245 by 3724 correct to four figures. The usual procedure is given below at the left.

3724)9376245(2517.78	3724)9376245(2518
7448	7448
19282	19282
18620	18620
6624	662
3724	372
29005	290
26068	296
29370	
26068	
33020	
29792	

Correct to four figures the quotient is 2518. The contracted form of the same division is shown above at the right, the quotient 2518 being obtained directly as follows:

The first two figures in the quotient, 25, are found in the usual way and the remainder is 662. It is not necessary to bring down any more figures of the dividend. The divisor may also be contracted (as we usually divide tentatively by the first one or two figures of the divisor anyhow). Thus cutting off the 4 the 372 is contained once in the remainder 662, the remainder being 290. Cutting off the 2 in the divisor, the 37 is contained in the 290 nearly 8 (more than $7\frac{1}{2}$) times. This 8 is the final desired (fourth) figure in the quotient.

19. Exercises and Problems.

In the following examples, determine the results correct to the stated number of decimal places by the methods explained and illustrated in articles 16, 17, and 18.

1. Find the sum of 23.45617, 937.34212, 42.31759, 532.23346, 141.423798, correct to two decimal places.
2. Subtract 987.642 from 993.624 correct to tenths.
3. Multiply 32.4736 by 24.7955 correct to five figures.
4. Divide 100 by 3.14159 correct to .01.
5. As will be seen later in detail, the circumference of a circle is found by multiplying the diameter of the circle by the number 3.14159. The diameter of a circle is measured as 16.23 inches. Calculate the circumference

correct to the nearest hundredth of an inch (the precision with which the diameter is measured).

6. The circumference of a round pole is measured as 26.4 inches, correct to the nearest tenth. What is the diameter, correct to the nearest tenth?

7. In insurance, annuity, pension and related calculations very many different amounts are calculated in United States money to several decimal places, and the totals or sums are then expressed in cents (to the nearest hundredth of a dollar). Express the sum of the following amounts correctly in cents: \$147.23689, \$5721.4658, \$81756.4732, \$21.66382.

8. The unit of measurement in the metric decimal system is the meter, which is 39.370 inches, correct to three decimal places. A length measured correctly to the nearest tenth is given as 486.7 inches. How many meters is this, expressed to the same precision?

9. Convert 32.59 meters to inches, correct to the nearest hundredth.

10. The area of a rectangle (oblong figure with straight edges and square corners) is found by multiplying the length by the width. The size of a rectangular sheet of paper is found by measuring as accurately as possible, to be 8.45 inches by 11.03 inches. What is the area, to the same accuracy, in square inches?

11. The volume of box or solid like a shoe-box or a brick is equal to the product of its length, width, and thickness or depth. How many cubic inches of clay are contained in a brick which measures 2.5 by 8 by 4.1 inches?

12. A certain cardboard box must hold one liquid quart which is exactly 57.75 cubic inches. The base is square and measures 3.25 inches on a side (width and thickness the same). What must be the length of the box, to the nearest hundredth of an inch?

Part 2

PRINCIPLES AND METHODS OF ARITHMETIC

Chapter 5

FACTORS, MULTIPLES AND DIVISORS

20. Definitions. If a certain number is the product of two or more other numbers, each of those other numbers is called a *factor* of that number. Thus, since $2 \times 3 = 6$, 2 and 3 are factors of 6. Similarly, the factors of 51 are 3 and 17; the factors of 30 are 2, 3, and 5; the factors of 81 are 3, 3, 3, and 3; etc.

A *multiple* of any number is the product obtained by multiplying that number by some other number. Thus, 42 is a multiple of 6, obtained by multiplying the 6 by 7.

A *divisor* has already been defined in article 13. In the sense in which we shall use it in this chapter, however, a *divisor* of any number is another number which is contained in that one a whole number of times, that is, exactly and without a remainder. Thus, 6 is a divisor of 42, the quotient being 7. Also, 3 is a divisor of 42, the quotient being 14. It is seen that a divisor is in reality a factor. This was to be expected, since, in division, the dividend is the product of the divisor and the quotient.

A common multiple of several numbers is a number which is a multiple of each of those numbers. Thus, 81 is a multiple of 3, 9, and 27. In particular, the *least common multiple* of several numbers is the smallest number which is a multiple of each of them. The Least Common Multiple is often denoted by the letters L.C.M. Thus, the L.C.M. of 2, 3, 4, and 5 is 60.

A common divisor of several numbers is a number which is a divisor of each of those numbers. Thus, 3 is a common divisor of 9, 18, 36, and 54. In particular, the *greatest common divisor* of several numbers is the largest number which is an exact divisor of each of them. The Greatest Common Divisor is often denoted by the letters G.C.D. Thus, the G.C.D. of the numbers 9, 18, 36, and 54 is 9.

A number which has other factors besides itself and 1 is called a *composite* number. One which has no other factors except itself and 1 is called a *prime* number. Thus, 14, having the factors 2 and 7 (besides 14 and 1), is a composite number, but 13, having no such factors (other than 13 and 1), is a prime number.

A number which has 2 as a factor, that is, is exactly divisible by 2, is called an *even* number. Other numbers, not divisible by 2, are *odd* numbers.

21. Tests for Certain Factors. In certain calculations in arithmetic, it is often of importance to be able to determine whether or not certain small numbers are factors of some given number. In the case of some numbers, this can be determined by rules which we shall now develop. In other cases the factors can be determined by tables based upon these rules.

Some of the following rules are very simple, and the reasons for them more or less obvious. Certain others are not so, and will be explained in some detail.

Since 2 and 5 are factors of 10, they are factors of any multiple of 10. Also, as any number can be considered as a multiple of 10 plus its last figure, we have at once the following rules:

(i) *If the last figure of any number is even, the number is divisible by* 2.

(ii) *If the last figure of any number is* 0 *or* 5 *the number is divisible by* 5.

Since 100 will contain 4 and 25 any multiple of 100 will contain 4 and 25. Also, any number is some multiple of 100 plus the number consisting of the last two figures. Therefore,

(iii) *If the number consisting of the last two figures of any given number is divisible by* 4 *or by* 25, *the entire number is divisible by* 4 *or* 25, *respectively.*

By a method similar to that just used for the factor 4 the following rule is also obtained:

(iv) *If the number represented by the last three figures of a given number is divisible by* 8, *the given number is divisible by* 8.

Consider next divisibility by 9. Since $10 = 9 + 1$, any number of 10's equals the same number of 9's plus the same number of 1's. Since $100 = 99 + 1$, any number of 100's equals the same number of 99's plus the same number of 1's; $1000 = 999 + 1$, and any number of 1000's equals the same number of 999's plus the same number of 1's; etc. Also 9, 99, 999, etc., are each multiples of 9. Therefore, any number is some multiple of 9 plus the sum of its separate figures, and so is divisible by 9 if the sum of its figures is. That is

(v) *If the sum of the separate figures of any number is divisible by* 9, *the number is divisible by* 9.

As an example, take 7362. Here $7 + 3 + 6 + 2 = 18$, which is divisible by 9. Therefore, 7362 is divisible by 9; the quotient is 818.

Now any number which contains 9 as a factor also contains 3 as a factor, and so if the sum of the figures of a number contains 9 it also contains 3. Therefore,

(vi) *If the sum of the figures of a number is divisible by* 3 *the number is divisible by* 3.

If any number is divisible by 2 and also by 3, it is divisible by $2 \times 3 = 6$. But if divisible by 2 it is even. Therefore,

(vii) *If any even number is divisible by* 3 *it is divisible by* 6.

By considering $10 = 11 - 1$, $100 = 99 + 1$, $999 = 1000 - 1$, $10000 = 9999 + 1$, and so on, alternating -1 and $+1$ for the alternate multiples of 10, it can be shown by a method similar to that used for (v) that

(viii) *If the sum of the even-placed figures of a number equals that of the odd-placed figures, or if the difference of the two sums is divisible by* 11, *the number is divisible by* 11.

For example, in the number 34267893 the odd-placed figures (1st, 3rd, 5th, etc.) are 3, 8, 6, 4 and their sum is 21. The even-placed figures (2nd, 4th, 6th, etc.) are 9, 7, 2, 3 and their sum is 21. These sums are equal and the number is therefore divisible by 11; the quotient is 3,115,263.

In the number 8172635482 the even-placed figures are 8, 5, 6, 7, 8 and their sum is 34; the odd-placed figures are 2, 4, 3, 2, 1 and their sum is 12. These sums are not equal but their difference, $34 - 12 = 22$, is divisible by 11. The given number is therefore divisible by 11; the quotient is 742,966,862.

The following rule is obvious:

(ix) *If the last figure of a number is a zero the number is divisible by* 10; *the quotient is found by striking off the zero.*

The rules (i) to (ix) provide tests for the factors 2, 3, 4, 5, 6, 8, 9, 10, 11; the test for 7 is too complicated to be useful and it is simpler to divide by 7 by short division to see if the quotient is exact.

The rules given can also be used to test for larger factors. For example, determine whether the number 12870 is divisible by 99. Testing by (v) the number is divisible by 9; the quotient is 1430. Testing 1430 by (viii) or by direct division, it is divisible by 11; the quotient is 130. The given number 12870 is therefore equal to $130 \times 11 \times 9$ or $130 \times (11 \times 9) = 130 \times 99$, and is therefore divisible by 99. This was found by successively testing for 9 and 11. Therefore, in general,

(x) *If a number is successively divisible by several other numbers it is divisible by their product.*

22. Tables of Factors, Multiples and Prime Numbers. There are no convenient rules for simply determining divisibility by numbers greater than 11, except by combining the rules for the smaller numbers, as in (x) of the preceding article. Also, there are no universal rules for determining whether any particular number is prime or composite (does not or does have factors). For these reasons many numbers have been tested by the rules of the preceding article and also by direct division in order to find whether or not they are prime or composite, and, if composite, to determine their factors. In modern times, this work has been greatly facilitated by the use of computing machines. The results of such tests have been incorporated in tables of various forms covering certain ranges of numbers. Tables I and II at the end of this book are such tables. We give here an explanation of these two tables.

Table I is both a multiplication table and a division table. If any number up to 400 can be found in the body of the table it is divisible by the number at the top of the column in which it is found and also by the first number at the left on the same line with the given number. Thus, 238 is divisible by 14 and by 17. Some numbers are found in the table in more than one place and so have more than two factors. Thus, 42 is divisible by 6 and 7 and also by 3 and 14; it is also divisible by 2 and 21 but these are not shown in the table.

Table II is divided into three parts and gives the prime numbers and also the smallest prime factors of composite numbers, up to 9600. On the first page of the table is given every third hundred beginning with zero (0, 300, 600, 900, etc.); on the second page, every third hundred beginning with 100 (100, 400, 700, etc.); and on the third page the remaining hundreds (200, 500, 800, etc.). The hundreds are given in the first column on each page and the tens and units on the first line at the top of each page. Only those numbers ending in 1, 3, 7, 9 are given (hundreds at left, tens and units at top) for the following reasons.

The rules in article 21 show that all numbers of more than one figure which end in 2, 4, 5, 6, 8, or 0 are composite. Those which end in 1, 3, 7, and 9 may or may not be composite. Since the rules provide tests for the factors of the numbers ending in 2, 4, 5, 6, 8, and 0, it is, therefore, only necessary to show in the table those which end in 1, 3, 7, and 9.

To find whether a certain number is prime or composite, and, if composite, to find its smallest (prime) factor, read the hundreds at the left and the last figures (added to the hundreds) at the top, and then look in the column with the last figures and on the line with the hundreds. If the space where this line and column meet contains a number, this number is the smallest (prime) factor of the given number, and the given number is a multiple of this number. If the space contains two dots, the given number is prime, that is, has no factors or divisors.

As an example, take the number 2747, which is on the first page of Table II. Finding 2700 at the left and looking on this line under 47 we find the number 41. The number 41 is therefore a factor of 2747, and also the smallest factor. Similarly, on the second page of the table, 7993 is found to have a blank, or two dots, on the line with 7900 under 93, and therefore has no factors. In the same way it can be found whether any other number less than 9600 is prime or composite, and if composite, what is its smallest factor. A brief explanation is given on each page of the table.

23. Greatest Common Divisor. It is often necessary in the study of fractions and in the solution of certain problems to know the greatest common divisor (G.C.D.) of two or more numbers. There are two general rules for finding the G.C.D. One is used when the factors of the two or more numbers are easily found by the rules of article 21 or the tables of article 22, the other when they are not.

The rule for G.C.D. of numbers easily factored is as follows:

(i) *Write the numbers on the same line and draw lines at the left and underneath as for short division.*

(ii) *Using short division, divide each number by any common divisor, writing each quotient below its dividend.*

(iii) *Divide the quotients by any common divisor in the same way.*

(iv) *Continue the repeated division until a set of quotients is obtained which have no common divisor.*

(v) *Multiply all the divisors together; the product is the required G.C.D. of the original numbers.*

As an illustration of the use of this rule we find the G.C.D. of the numbers 28, 140, and 420. To begin with, a common divisor of all three is 7.

7)	28	140	420
4)	4	20	60
	1	5	15

The numbers 1, 5, 15 have no common divisor (except 1). Therefore, the G.C.D. of 28, 140, and 420 is the product of the divisors, $7 \times 4 = 28$.

The rule for finding the G.C.D. of numbers not readily separated into factors is as follows:

(i) *Draw two vertical lines and write down the smaller of the two given numbers outside the two lines, the greater of the two one line above the level of the other.*

(ii) *Divide the greater of these two numbers by the smaller, writing the quotient between the vertical lines, the product of the quotient and divisor under the dividend, and the remainder below the product after subtraction.*

(iii) *Divide the smaller of the two original numbers by the remainder or vice versa, using the smaller of the two for divisor, and find a new remainder as in* (ii).

(iv) *Divide the last divisor by the last remainder, and repeat the process until there is no remainder. The last divisor is the G.C.D. of the two original numbers.*

(v) *Take this G.C.D. and another of the originally given numbers and in the same manner find their G.C.D. Repeat this procedure until all the original given numbers have been used.*

(vi) *The last G.C.D. found is the required G.C.D. of all the given numbers.*

As an illustration of the application of this rule we find the G.C.D. of 18607 and 417979. The calculation is carried out at the left. The last divisor is 23, which is contained in 92 exactly 4 times, leaving no remainder. The G.C.D. of 18,607 and 417,979 is therefore 23.

		417979
18607	2	37214
		45839
	2	37214
17250	2	8625
1357	6	8142
966	2	483
391	1	391
368	4	92
23	4	92

Had there been more numbers than these two given, this 23 would be taken with another of the given numbers and their G.C.D. found in the same way, and so on until all had been used.

24. Least Common Multiple. The least common multiple (L.C.M.) of several numbers may be found by the following rule:

(i) *By the rules of article* 21 *or the tables of article* 22 *find the prime factors of each of the numbers.*

(ii) *Find the product of the prime factors of the largest number, and also that of such prime factors of the other numbers as are not contained in the largest. (If any factor appears more than once in any numbers it is to be included in this second product as if the repeated factors were different.)*

(iii) *The product of these two products is the L.C.M. of the given numbers.*

We now apply this rule to find the L.C.M. of the three numbers 63, 66, 78.

$$\text{(i) Prime factors: } \begin{cases} 63 = 3 \times 3 \times 7 \\ 66 = 2 \times 3 \times 11 \\ 78 = 2 \times 3 \times 13 \end{cases}$$

(ii) Product of prime factors of 78 is $2 \times 3 \times 13 = 78$; product of *other* prime factors of 66 and 63 is $3 \times 7 \times 11 = 231$. (Here one of the two equal factors 3 in 63 is counted as another factor.)

(iii) L.C.M. $= 78 \times 231 = 18{,}018$.

The following rule for L.C.M. is also useful; it is somewhat longer than the one already given but is more nearly automatic and easier to use.

(i) *Write the given numbers in a line, omitting such of the smaller numbers as are themselves factors of the larger.*

(ii) *Using short division divide by any prime factor which may be contained in one or more of these numbers, writing the quotients and undivided numbers below the dividends, omitting the* 1's.

(iii) *Using these quotients and undivided numbers as new dividends, divide again by some prime factor, and continue the process until there is no common prime factor of the quotients.*

(iv) *Find the product of* ALL *the divisors and the* FINAL *quotients. This product is the required L.C.M.*

We now apply this rule to find the L.C.M. of the numbers 64, 84, 96, and 216.

2)	64	84	96	216
2)	32	42	48	108
3)	16	21	24	54
2)	16	7	8	18
2)	8	7	4	9
2)	4	7	2	9
	2	7	1	9

The L.C.M. is $2 \times 2 \times 3 \times 2 \times 2 \times 2 \times 2 \times 7 \times 9 = 12{,}096$.

25. Use of G.C.D. and L.C.M. in Solving Problems. Many very common problems require for their solution the finding of the G.C.D. or the L.C.M. of two or more numbers. Such problems are usually very simple in principle, but when solved by trial methods they may seem very difficult. For the purpose of illustration a few problems of this type will be solved here.

Problem 1. A farmer wishes to put 364 bushels of corn, 455 bushels of oats and 546 bushels of wheat in the least possible number of bins of the same size without mixing, filling each bin; what number of bushels must each bin hold?

Solution. It is seen at once that to fill each bin and avoid mixing the different grains, the capacity of a bin must be an exact divisor of the number of bushels of *each* grain, and to be the largest possible the capacity in bushels must be the *greatest common* divisor. We, therefore, have to find the G.C.D. of 364, 455, and 546. By the rules of article 23 this is found to be 91. The capacity of each bin must therefore be 91 bushels.

Problem 2. Rails of equal length are to be nailed end to end on posts to form a straight fence 7 rails high enclosing a field 14,599 by 10,361 feet, and it is desired to use the longest rails possible without cutting. How long must the rails be and how many will be needed?

Solution. In order to avoid cutting, the length of rail must be a divisor of the length and of the width of the field, and to use the longest rail possible the length must be the *greatest* divisor. We have, therefore, to find the G.C.D. of 10,361 and 14,599. Using the rule of article 23 the G.C.D. is found to be 13. The rails must therefore be 13 feet long.

On the long side of the field will be used $14{,}599 \div 13 = 1123$ rails, and on the short side $10{,}361 \div 13 = 797$ rails. For a 7-rail fence on all four sides will be required $7 \times (2 \times 1123 + 2 \times 797) = 7 \times (2246 + 1594) = 7 \times 3840 = 26{,}880$ rails.

Problem 3. A earns \$620, B earns \$1116, and C \$1488 in a whole number of weeks, all working for the same weekly salary. What is the most that salary can be and how many weeks has each worked to earn the amount named?

Solution. Since the different amounts are earned each in a whole number of weeks and since each man draws the same in each week, the weekly salary is a divisor of each of the amounts named, and for it to be the largest possible it must be the G.C.D. of the three amounts. To find the G.C.D. by the rule of article 23, we first find that of the two smaller numbers, 620 and 1116; it is 124. We then find the G.C.D. of this 124 and the other number, 1488; this is also 124. The G.C.D. of 620, 1116, and 1488 is, therefore, 124 and this is the common monthly salary in dollars.

The weeks required for A to earn the amount stated are then $620 \div 124 = 5$; for B, $1116 \div 124 = 9$; and for C, $1488 \div 124 = 12$.

Problem 4. If A can build 14 rods of fence in a day, B can build 25 rods, C 8 rods and D 20 rods, what is the least number of rods of fence that will furnish a number of whole day's work for all of the four?

Solution. Since the total amount is to furnish a number of whole day's work for all, it must be an exact multiple of the amount per day that each can do, and since it is to be the least possible, it must be the *least common* multiple of the four daily amounts. We have, therefore, to find the L.C.M. of 8, 14, 20, and 25. Using either of the rules of article 24 the L.C.M. is found to be 1400 rods. This will furnish 100 days' work for A, 56 days for B, 175 for C, and 70 for D.

Problem 5. The front, middle and rear wheels of a truck and trailer are respectively 10, 12, and 15 feet in circumference, and a rivet in the rim of each is at the ground when the truck starts. How far must it travel before the same three rivets are again at the ground at the same time?

Solution. For the same three rivets to reach the ground at the same time again, each wheel must turn over a whole number of times, and the distance traveled must therefore be a *multiple* of the circumference of each; also for the *first* time they all reach the ground together again the distance must be the *least* in which that can happen. We have, therefore, to find the L.C.M. of 10, 12, and 15. By the rule of article 24 the L.C.M. is 60. The same three rivets will therefore reach the ground together again 60 feet from the starting point.

26. Exercises and Problems.

1. Find without dividing whether or not 2, 4, and 8 are factors of the following numbers; after testing by the rules verify by actual division: 23576; 193888; 7632458.

2. In the same way test the following for divisibility by 5, 10, and 25 and verify: 73565; 73575; 823450.

3. In the same way test and verify the following for divisibility by 3, 6, and 9: 423966; 1894869; 7245472.

4. Find the G.C.D. of 2041 and 8476.
5. Find the G.C.D. of 336 and 812.
6. Find the G.C.D. of 4718, 6951, and 8876.
7. Find the L.C.M. of 14, 19, 38, 42, and 57.
8. Find the L.C.M. of 9, 15, 25, 35, 45, and 100.
9. Find the L.C.M. of 6, 8, 10, 15, 18, 20, and 24.

10. A three-sided plot of land 165, 231, and 385 feet on a side is to be enclosed with a fence having panels of the greatest possible uniform length; what must be the length of each panel?

11. What is the smallest sum of money with which can be purchased either table radios at $49.00 apiece, electric razors at $21.00, or phonographs at $72.00?

12. The front wheel of a truck is 11 feet in circumference and the rear wheel 15 feet, and a certain spoke in each points upward as the truck starts. How far must the truck travel before the same spokes point upward together again for the 575th time?

13. Find from Table II whether the following numbers are prime or composite, and if composite find their least factors: 41, 91, 169, 193, 649, 983, 2249, 3551, 6551, 9421, 9599.

14. The troop transport planes in a certain group can carry respectively 30, 33, 42, 45 men. What is the smallest number of men which can be carried by the group, each plane being fully loaded and each making one or more trips?

15. In a row of electric fans lined up on a shelf, some have 2 fan blades, some have 3 blades, some 4, some 5, and some 6. If all start running at the same instant with one blade on each pointing directly upward, how many revolutions will those of each kind make before the same blade on each as at the start will point upward at the same instant again for the fifth time after the start?

16. A customer plans to buy articles at a store which cost respectively 15, 18, 20, 21, 35, 42 cents. What is the smallest sum of money with which he may buy selections of one or more articles of each of the listed prices?

17. At registration time at a college the numbers enrolled for various courses are: 84, 112, 140, 28, 252, 56, 420, 336, and 728. If all classes formed in the various courses are to be of the same size, what is the largest size class that may be formed, and how many classes are there in all?

18. In decorating the borders of circular frames, an artist divides the outer edge into sections of equal length and paints the sections in different colors. What is the length of the longest section that can be used for a set of circles having circumferences of 276, 84, 168, 132, 72, 156, 324, and 108 inches?

Chapter 6

FRACTIONS

27. Definitions. By way of review, a few of the definitions connected with fractions have in article 10 been recalled from elementary arithmetic, but as the properties and handling of fractions are now to be considered in more detail, we will repeat some of the definitions.

If a single unit, as 1 foot, 1 pound, 1 object of any kind, be divided into equal parts each of the parts is called a *fractional unit* and is named according to the number of equal parts. Thus, if the unit is divided into four equal parts, each fractional unit is called *one fourth.* This is written in symbols $\frac{1}{4}$. Similarly, each of ten equal parts is called one tenth, $\frac{1}{10}$; etc.

A *fraction* is one or the sum of several fractional units. Thus $\frac{1}{4}$ is a fraction; so, also, is three fourths, $\frac{1}{4} + \frac{1}{4} + \frac{1}{4}$, which is written $\frac{3}{4}$. The symbol $\frac{3}{4}$ indicates 3 of 4 equal parts, three of the fractional units $\frac{1}{4}$. Similarly, $\frac{2}{3}$ is two thirds; $\frac{9}{11}$ is nine elevenths; and so on.

In a fraction such as $\frac{1}{4}$, $\frac{1}{2}$, $\frac{3}{4}$, $\frac{2}{3}$, $\frac{9}{11}$, etc., the horizontal line separating the two numbers in each fraction is called the *fraction line.* The number above the fraction line is the *numerator* and that below is the *denominator*, of the fraction. The denominator *names* the fractional unit and the numerator indicates the number of those units contained in the fraction. Thus, in $\frac{1}{2}$ the 1 is the numerator and the 2 is the denominator; in $\frac{3}{4}$ the numerator is 3 and the denominator is 4. The numerator and denominator taken together are sometimes called the *terms* of the fraction.

A fraction whose numerator is less than its denominator is called a *proper* fraction; one whose numerator is equal to or greater than its denominator is an *improper* fraction. Thus, $\frac{2}{3}$, $\frac{9}{11}$ are proper fractions; $\frac{7}{7}$, $\frac{81}{72}$, $\frac{4}{3}$ are improper fractions.

A number which is expressed by a whole number and a fraction is called a *mixed number.* Thus, $12\frac{3}{4}$ is a mixed number; it is read "twelve and three fourths."

Fractions as defined in this article are called *common fractions* to distinguish them from decimal fractions, discussed in article 10.

28. Values and General Principles of Fractions. If a unit be divided into three equal parts, each part is larger than one of four equal parts of the same unit. That is, $\frac{1}{3}$ is greater than $\frac{1}{4}$; similarly, $\frac{1}{2}$ is greater than $\frac{1}{3}$, and $\frac{1}{4}$ is greater than $\frac{1}{5}$. Also $\frac{2}{3}$ is greater than $\frac{2}{5}$, and $\frac{3}{7}$ is greater than $\frac{3}{9}$. In general, when the numerators of different fractions are the same, the fractions having *larger* denominators have *smaller* values.

On the other hand, two thirds are obviously more than one third; four fifths are more than two fifths; three fourths more than one fourth. That is, $\frac{2}{3}$ is greater than $\frac{1}{3}$; $\frac{4}{5}$ is greater than $\frac{2}{5}$; $\frac{3}{4}$ is greater than $\frac{1}{4}$; etc. In general, when the denominators are the same, fractions having *larger* numerators have *larger* values. When the denominators and the numerators of different fractions are both different, the values of the fractions may be compared by converting them so that they have the same denominators. We shall see in article 30 how this is done.

If one unit is divided into, for example, thirds, the fractional unit is $\frac{1}{3}$, and this is the third part of 1, that is, $1 \div 3$. Similarly, $\frac{2}{3}$ is two thirds of one unit, or one third of two units, that is, $2 \div 3$. Similarly, $\frac{3}{4}$ indicates $3 \div 4$; $\frac{1}{2}$ indicates $1 \div 2$; $\frac{17}{23}$ means $17 \div 23$; etc. In general, *fractions indicate division*, the numerator being a dividend, the denominator a divisor, and the *value* of the fraction the quotient.

Since fractions indicate division, all changes in the terms of a fraction (numerator and denominator) will affect its value (quotient) according to the general principles of division. These relations constitute the *general principles of fractions* and will be utilized in the following articles to develop the rules for the usual arithmetical operations with fractions.

29. Reduction of Fractions. If a unit is divided into four equal parts (fourths), and each fourth is then divided into three equal parts, there will then be a total of twelve equal parts, and each of these is one twelfth. Thus, each fourth contains three twelfths. That is, one fourth *equals* three twelfths. This is written $\frac{1}{4} = \frac{3}{12}$, or $\frac{3}{12} = \frac{1}{4}$. This means that the two fractions $\frac{3}{12}$ and $\frac{1}{4}$ have the same *value* even though they are of different *forms*. The fraction $\frac{1}{4}$ is said to be expressed in *lower terms* than the equivalent or equal fraction $\frac{3}{12}$. Similarly, $\frac{8}{12} = \frac{2}{3}$, and the fractions $\frac{3}{12}$ or $\frac{8}{12}$, when written as equivalent to $\frac{1}{4}$ or $\frac{2}{3}$, respectively, are said to be *reduced to lower terms*.

On examination it is seen that in order to reduce $\frac{3}{12}$ to $\frac{1}{4}$ the numerator and denominator are each separately divided by 3. Similarly, if numerator and denominator of $\frac{8}{12}$ be each separately divided by 4, $\frac{8}{12}$ is reduced to the equivalent fraction $\frac{2}{3}$. In the same way $\frac{42}{54}$ is reduced to $\frac{7}{9}$ by dividing numerator and denominator by 6.

In general, a fraction can be reduced to lower terms if numerator and denominator are divisible by a single number, that is, *if they have a common divisor*. In order to reduce a fraction to its *lowest* terms, therefore, it is seen at once that the *greatest* common divisor must be used. Thus, we have the following

RULE: *To reduce a fraction to its lowest terms, divide its numerator and its denominator by their G.C.D.*

If the numerator and denominator of a fraction are small numbers, the G.C.D. can usually be found by simple inspection, otherwise the rule for finding the G.C.D. must be used (article 23).

To reduce $\frac{18}{30}$ to its lowest terms it is seen at once that the G.C.D. of 18 and 30 is 6. The numerator of the reduced fraction is, therefore, $18 \div 6 = 3$, and the denominator is $30 \div 6 = 5$; the reduced fraction is, therefore, $\frac{3}{5}$. To reduce $\frac{336}{812}$ to its lowest terms, we have (as found in Ex. 5, art. 26) for the G.C.D. of 336 and 812, 28. The reduced fraction is, therefore, $\frac{(336 \div 28)}{(812 \div 28)} = \frac{12}{29}$.

Since any fraction can be reduced to lower terms by *dividing* numerator and denominator by their G.C.D. (if they have one), then any fraction can be raised to higher terms by *multiplying* its numerator and denominator both by a single number. A use for this operation will be seen presently.

We have in the preceding discussion developed the idea and the process of reducing fractions to higher or lower terms by proceeding in a simple manner from the definition of a fraction. We have seen, however (in the preceding article), that the general principles of fractions are the same as those for ordinary division. If, therefore, the numerator and denominator are both multiplied or divided by the same number, the quotient (the value of the fraction) will remain unchanged. From the general principles of fractions, therefore, we could have derived the reduction rules directly.

Since a fraction indicates a division (article 28) the value of an improper fraction will be greater than 1, because the numerator is greater than the denominator. Thus, $\frac{15}{5} = 15 \div 5 = 3$. This means that fifteen fifths is equivalent to three units, each unit being equivalent to five fifths. Similarly, $\frac{17}{5} = 17 \div 5 = 3\frac{2}{5}$, according to the rule of article 9 for division with a remainder; $\frac{15}{4} = 15 \div 4 = 3\frac{3}{4}$; $\frac{25}{2} = 12\frac{1}{2}$; $\frac{80}{15} = 80 \div 15 = 5\frac{5}{15} = 5\frac{1}{3}$, or, first reducing to lower terms, $\frac{80}{15} = \frac{16}{3} = 5\frac{1}{3}$.

In general, an improper fraction can be reduced to a whole or a mixed number by the ordinary rule for division, as follows:

RULE: *Divide numerator by denominator, first reducing to lower terms, if possible, and if there is a remainder, write it over the dividing denominator. The quotient or the quotient and the remainder fraction form the equivalent reduced value of the improper fraction.*

A mixed number may be expressed as an improper fraction by simply reversing this rule. This gives the following

RULE: *Multiply the whole number part of the mixed number by the denominator of the fractional part and to the product add the numerator of the fractional part; write the sum as a numerator over the original denominator, to form the desired fraction.*

Example: $7\frac{3}{4} = \dfrac{(4 \times 7) + 3}{4} = \dfrac{28 + 3}{4} = \dfrac{31}{4}.$

30. Fractions of Common Denominator. We saw in article 29 that in order to compare the values of fractions, they should have the same denominator. When two or more fractions have the same number for denominators, this number is called the *common denominator* of the several fractions. Thus, in the fractions $\frac{3}{8}$, $\frac{7}{8}$, $\frac{5}{8}$, 8 is the common denominator.

The fractions $\frac{3}{4}$ and $\frac{5}{8}$ do not have a common denominator, but by raising $\frac{3}{4}$ to higher terms, it can be expressed with denominator 8. Thus, if numerator and denominator of $\frac{3}{4}$ be multiplied by 2 it becomes $\frac{6}{8}$, which has the same denominator as $\frac{5}{8}$.

Consider the fractions $\frac{2}{3}$ and $\frac{3}{5}$. They do not have a common denominator and neither can be expressed with the denominator of the other by raising or reducing its terms. Both map, however, be expressed with the denominator 15, or 30, etc. Thus, if numerator and denominator of $\frac{2}{3}$ be multiplied by 5 it becomes $\frac{10}{15}$, and if numerator and denominator of $\frac{3}{5}$ be multiplied by 3 it becomes $\frac{9}{15}$; and in these forms the two have the common denominator 15. Similarly, they may be written as $\frac{20}{30}$ and $\frac{18}{30}$, $\frac{30}{45}$, and $\frac{27}{45}$, etc., the least number which will be a common denominator of $\frac{2}{3}$ and $\frac{3}{5}$, however, is 15. This is the *least common denominator*, or the *lowest* common denominator of $\frac{2}{3}$ and $\frac{3}{5}$. This is sometimes denoted by the letters L.C.D.

The number 15 is the least number which is an exact multiple of 3 and 5 and is, therefore, the *least common multiple* of the denominators 3 and 5. Similarly, the least common denominator of the fractions $\frac{5}{6}$, $\frac{5}{9}$, $\frac{11}{36}$ is 36, the L.C.M. of the denominators 6, 9, 36. In general, *the L.C.D. of two or more fractions is the L.C.M. of their several denominators.* If the denominators are such that their L.C.M. is not readily found by inspection, it is to be found by the rules of article 24.

Any set of fractions may be expressed with a common denominator as follows:

(i) *Find the L.C.M. of the denominators; this is the L.C.D. of the fractions.*

(ii) *Divide the L.C.D. by each individual denominator and multiply the quotient by the corresponding numerator, to obtain the new numerator.*

(iii) *Write each new numerator over the common denominator; the resulting fractions are the original fractions expressed with the L.C.D.*

31. Addition and Subtraction of Fractions. Since $\frac{3}{7}$ and $\frac{2}{7}$ are expressed in the same fractional unit (sevenths), they may be compared or combined. Thus, in the same way as we would say that three dollars plus two dollars is equal to five *dollars*, we would say three sevenths plus two sevenths is equal to five *sevenths*. This is written $\frac{3}{7} + \frac{2}{7} = \frac{5}{7}$. Similarly, $\frac{7}{8} + \frac{5}{8} = \frac{12}{8} = 1\frac{4}{8} = 1\frac{1}{2}$; $\frac{2}{12} + \frac{5}{12} + \frac{1}{12} = \frac{8}{12} = \frac{2}{3}$; $\frac{4}{19} + \frac{3}{19} + \frac{7}{19} + \frac{2}{19} = \frac{16}{19}$; etc. From these examples it is at once obvious that when fractions have a common denominator, they can be added by simply *adding the numerators* and writing the sum over the same denominator.

In the same way it is seen at once that $\frac{3}{7} - \frac{2}{7} = \frac{1}{7}$; $\frac{7}{8} - \frac{5}{8} = \frac{2}{8} = \frac{1}{4}$; $\frac{16}{19} - \frac{7}{19} = \frac{9}{19}$; etc., and, in general, any fractions with a common denominator are subtracted by *subtracting he numerator* of the subtrahend fraction from that of the minuend fraction, and writing the remainder over the common denominator to form the remainder fraction.

If fractions which are to be added or subtracted do not have the same denominator, they may be expressed with a common denominator by the method of the preceding article, and then added or subtracted. We have, therefore, for the addition and subtraction of fractions the following

RULE: *To add or subtract fractions, first express them with the L.C.D., and then add or subtract the numerators, writing the result as the numerator of a fraction with the common denominator. This fraction is the desired sum or difference, respectively.*

Example: To add $\frac{2}{3} + \frac{5}{8} + \frac{5}{6} + \frac{3}{4}$. The L.C.M. of 3, 8, 6, and 4 is 24 and the fractions are $\frac{16}{24}$, $\frac{15}{24}$, $\frac{20}{24}$, $\frac{18}{24}$. The sum of the numerators is now $16 + 15 + 20 + 18 = 69$, and the sum of the fractions is, therefore, $\frac{69}{24}$. Reduced to lower terms this is $\frac{23}{8}$ and expressed as a mixed number it is $2\frac{7}{8}$. The sum of the given fractions is therefore $2\frac{7}{8}$.

To subtract: $\frac{17}{21} - \frac{2}{9}$; the L.C.D. is 63 and the subtraction becomes $\frac{51}{63} - \frac{14}{63} = \frac{37}{63}$.

To add or subtract mixed numbers, the whole number parts are added or subtracted in the usual manner, and the fractional parts are added or subtracted according to the rule just given and illustrated. If the fractional part of the subtrahend is greater than that of the minuend, one unit is "borrowed" from the whole number part of the minuend as in ordinary subtraction, and added to the fractional part of the minuend. This 1 and fraction are then expressed as an improper fraction and the fractional part of the subtrahend subtracted as before. In subtracting the whole number parts it must then be remembered that the minuend has been decreased by 1. These operations are illustrated in the three following examples:

$$6\tfrac{7}{8} - 4\tfrac{5}{8} = 2\tfrac{2}{8} = 2\tfrac{1}{4}.$$

$$15\tfrac{2}{3} - 11\tfrac{1}{5} = 15\tfrac{10}{15} - 11\tfrac{3}{15} = 4\tfrac{7}{15}.$$

$$9\tfrac{2}{9} - 3\tfrac{17}{27} = 8\tfrac{11}{9} - 3\tfrac{17}{27} = 8\tfrac{33}{27} - 3\tfrac{17}{27} = 5\tfrac{16}{27}.$$

32. Multiplication of Fractions. Six sevenths is obviously twice as great as three sevenths; that is, two times three sevenths is six sevenths. This is written $2 \times \frac{3}{7} = \frac{6}{7}$. Similarly, $3 \times \frac{1}{4} = \frac{3}{4}$; $5 \times \frac{7}{8} = \frac{35}{8}$; etc. In these illustrations it is seen that the product is a fraction with a denominator which is the same as that of the multiplicand fraction, and a numerator which is the product of the multiplier and the numerator of the multiplicand. This is in accordance with the principle of article 28, increasing the numerator increases the value of the fraction; multiplying the *numerator* multiplies the fraction. From this we have the

RULE: *To multiply a fraction by a whole number, multiply the numerator by that number, and write the product as the numerator of a new fraction with the same denominator. This fraction is the desired product.*

Thus, to multiply $\frac{17}{28}$ by 3 we have $3 \times \dfrac{17}{28} = \dfrac{3 \times 17}{28} = \dfrac{51}{28}$.

From the same general principles of fractions, or of division of numbers, as stated in article 28, it is also seen at once that, since increasing the denominator (divisor) decreases the fraction (quotient), then multiplying the denominator is the same as dividing the fraction. That is, in order to divide a fraction by any number, multiply the denominator by that number. This is at once expressed in the following

RULE: *To divide a fraction by a whole number, multiply the denominator of the fraction by that number and write the product as the denominator of a new fraction with the same numerator.*

Example: $\frac{3}{4} \div 7 = \frac{3}{7 \times 4} = \frac{3}{28}.$

In order to apply either of the last two rules to a mixed number, multiply or divide the whole number and fractional parts separately and add the results, or first reduce the mixed number to an improper fraction and then apply the rule directly.

Since division by, say, 3 is to take one third of the dividend, then division by 3 is the same as multiplication by $\frac{1}{3}$, for $\frac{1}{3} = 1 \div 3$ and to multiply by 1 and then divide by 3 is simply to divide by 3. To multiply a fraction by the fraction $\frac{1}{3}$ is, therefore, to divide the fraction multiplicand by 3, that is, multiply its denominator by 3. In general, therefore, we have the

RULE: *To multiply any fraction by another fraction having the numerator* 1, *multiply the denominator of the multiplicand fraction by the denominator of the multiplier fraction and write the product as the denominator of a new fraction with the numerator of the original multiplicand as its numerator.*

Examples: $\frac{3}{4} \times \frac{1}{2} = \frac{3}{2 \times 4} = \frac{3}{8}$; $\frac{15}{28} \times \frac{1}{6} = \frac{15}{168} = \frac{5}{56}$; etc.

We next consider the multiplication of any two fractions, having any numerators and denominators. Thus, suppose we have to find the product $\frac{2}{3} \times \frac{4}{5}$. Since $\frac{4}{5}$ is the same as $4 \times \frac{1}{5}$ this is the same as $\frac{2}{3} \times 4 \times \frac{1}{5}$. That is, multiplication by the fraction $\frac{4}{5}$ is the same as multiplication by 4 and by $\frac{1}{5}$. But in order to multiply the fraction $\frac{2}{3}$ by 4 we multiply the numerator 2 by 4, and in order to multiply $\frac{2}{3}$ by $\frac{1}{5}$ we multiply the denominator 3 by 5. The net result of the full multiplication is, therefore, that we multiply the numerators of the two fractions together, 2×4, and multiply the two denominators together, 3×5. This is written $\frac{2}{3} \times \frac{4}{5} = \frac{2 \times 4}{3 \times 5} = \frac{8}{15}.$

In the same way the product of the two fractions $\frac{12}{17}$ and $\frac{4}{5}$ is $\frac{12}{17} \times \frac{4}{5} = \frac{12 \times 4}{17 \times 5} = \frac{48}{85}$; and similarly, $\frac{7}{9} \times \frac{11}{9} = \frac{7 \times 11}{9 \times 9} = \frac{77}{81}$; etc. In general, for the multiplication of *any* number of fractions, we have the very important

RULE: *The product of fractions is found by multiplying together the numerators, and also the denominators. The product of the numerators is the numerator of the product fraction, and the product of the denominators is the denominator of the product fraction.*

The use of this rule is illustrated by the following examples:

$$\frac{5}{11} \times \frac{2}{3} = \frac{5 \times 2}{11 \times 3} = \frac{10}{33}; \quad \frac{2}{3} \times \frac{5}{7} \times \frac{9}{11} = \frac{2 \times 5 \times 9}{3 \times 7 \times 11} = \frac{90}{231} = \frac{30}{77};$$

$$\frac{7}{8} \times \frac{9}{10} \times \frac{1}{4} \times \frac{2}{5} = \frac{126}{1600} = \frac{63}{800}; \quad \frac{2}{3} \times \frac{3}{5} = \frac{2 \times 3}{3 \times 5} = \frac{6}{15} = \frac{2}{5}.$$

Consider in particular the last of these examples. Here it is seen that the product $\frac{2}{5}$ is the same as the product fraction $\frac{2 \times 3}{3 \times 5}$ with the 3's in numerator and denominator removed or *cancelled* out. The entire operation can then be indicated as

$$\frac{2}{3} \times \frac{3}{5} = \frac{2 \times \not{3}}{\not{3} \times 5} = \frac{2}{5} \text{ or as } \frac{2}{\not{3}} \times \frac{\not{3}}{5} = \frac{2}{5}.$$

In the same way, the multiplication $\frac{5}{8} \times \frac{24}{30} \times \frac{7}{10}$ can be written out in the form

$$\frac{5}{8} \times \frac{24}{30} \times \frac{7}{10} = \frac{\overset{\not{3}}{\not{5} \times \not{2}\not{4} \times 7}}{\underset{\underset{2}{\not{6}}}{\not{8} \times \not{3}\not{0} \times 10}} = \frac{7}{2 \times 10} = \frac{7}{20}$$

Here the 8 is divided into the 24 and the quotient is 3; the 5 into the 30 and the quotient is 6; and finally the 3 into the 6, quotient 2. The remaining factors in the denominator are then 2×10 and in the numerator 7, so that the product-fraction becomes $\frac{7}{2 \times 10}$ or $\frac{7}{20}$, which is the product of the three original fractions.

In each of the preceding illustrations it is to be noted that factors (multipliers) which occur in *both* numerator and denominator of the product-fraction are cancelled or divided out. This simplifying process in the multiplication of fractions is called *cancellation*. In general, when the same factor occurs in both numerator and denominator of the product-fraction, or the same number occurs in a numerator *and* a denominator of any of the fractions being multiplied, it may be cancelled. This is at once seen to be the same in principle as dividing numerator and denominator by the same divisor to reduce a fraction to lower terms.

As a final illustration let us find by cancellation the value of the following expression:

$$\frac{2 \times 5 \times 16 \times 30 \times 15}{4 \times 8 \times 10 \times 7 \times 3 \times 25}.$$

This is carried out as follows:

$$\frac{\cancel{2} \times \cancel{5} \times \overset{\cancel{2}}{\cancel{16}} \times \overset{\cancel{3}}{\cancel{30}} \times \overset{③}{\cancel{15}}}{\underset{\cancel{2}}{\cancel{4}} \times \cancel{8} \times \cancel{10} \times ⑦ \times \cancel{3} \times \underset{\cancel{5}}{\cancel{25}}} = \frac{3}{7}$$

Cancelling the 2 in 4, the 8 in 16, the resulting 2's, the 10 in 30, the resulting 3 and 3, the 5 in 25 and the resulting 5 in 15, there remains in the numerator the factor 3 and in the denominator the factor 7. The reduced value of the original expression is, therefore, $\frac{3}{7}$.

In multiplication of fractions, either by other fractions or by whole numbers, all common factors in numerator and denominator should be cancelled before writing out the product. For example,

$$4 \times \frac{3}{16} = \frac{\cancel{4} \times 3}{\underset{4}{\cancel{16}}} = \frac{3}{4}.$$

33. Division of Fractions. Although the preceding article is devoted to the multiplication of fractions, it was seen that because of the general principles of fractions, multiplication and division of a fraction by a whole number are so closely related that we there obtained directly the

RULE: *To divide a fraction by a whole number, multiply the denominator of the fraction by that number.*

This rule has already been illustrated, thus: $\frac{5}{7} \div 2 = \frac{5}{7 \times 2} = \frac{5}{14}$; etc.

From this rule, as well as in ordinary multiplication of whole numbers, it is seen that division is exactly the inverse operation to multiplication. Thus to divide by 3, or $\frac{3}{1}$, is to multiply by $\frac{1}{3}$. Similarly, to divide by $\frac{1}{3}$ is to multiply by $\frac{3}{1}$, or 3. Also, to *multiply* by $\frac{2}{3}$ is to multiply by 2 and divide by 3. To *divide* by $\frac{2}{3}$, the inverse operation, is, therefore, to multiply by 3 and divide by 2, which is the same as multiplying by $\frac{3}{2}$. Similarly, to divide by $\frac{7}{8}$ would be to multiply by $\frac{8}{7}$, and in general, *to divide by any fraction is to multiply by that fraction inverted.* Thus, $15 \div \frac{7}{8} = 15 \times \frac{8}{7} = \frac{120}{7}$. We can therefore write as a

RULE: *To divide a whole number or a fraction by a fraction, invert the divisor fraction and multiply.*

In this multiplication after inversion of the divisor, the usual rules for multiplication, as developed in the preceding article, always apply.

As an example of this rule, to divide the fraction $\frac{15}{16}$ by the fraction $\frac{5}{8}$ we have $\frac{15}{16} \div \frac{5}{8} = \frac{15}{16} \times \frac{8}{5}$. In carrying out the final multiplication cancellation is to be used wherever possible. In this case we have

$$\frac{\overset{3}{\cancel{15}}}{\underset{2}{\cancel{16}}} \times \frac{\cancel{8}}{\cancel{5}} = \frac{3}{2}.$$

34. Conversion of Common Fractions into Decimals and Vice Versa. The meaning of a *decimal fraction* and of a *decimal number* are explained in article 10, and we come now to the relation between common fractions and decimal fractions.

As seen in article 10, a decimal fraction is simply a fraction whose denominator is 10, 100, 1000, 10000, etc., and is written without the denominator, the decimal point taking the place of the fraction line used in common fractions, and the position of the numerator figures in the first, second, third, etc., place to the right of the decimal point indicating, respectively, tenths, hundredths, thousandths, etc.

In division of decimals it was seen that a smaller number may be divided by a larger, the quotient being a decimal fraction. Thus, by carrying out the division in the usual manner by long division, it is found that $5 \div 8 = .625$. But, since a common fraction indicates division, the numerator being the dividend and the denominator the divisor, the fraction $\frac{5}{8}$ means the same thing as $5 \div 8$. That is, $\frac{5}{8} = .625$, which states that the *common fraction is equivalent to a decimal fraction.* The same is true of any common fraction. This means that any common fraction can be expressed as, or converted into, a decimal fraction. Since any common fraction is an indicated division, the method of conversion is plain: simply *divide the numerator by the denominator* by long division and follow the usual rules for decimal points. In order to carry out the process, therefore,

RULE I: *Write the numerator of a common fraction, followed by a decimal point and several ciphers, and divide it by the denominator, using long division and continuing to as many decimal places as desired or required. The quotient is the equivalent decimal fraction.*

If the quotient comes out exact, the decimal fraction is the exact equivalent of the common fraction; if not, the common fraction has no exact decimal equivalent. It may be expressed decimally as closely as desired, however, by carrying the division to a sufficient number of decimal places. As an illustration, let us find the decimal equivalent of the common fraction $\frac{13}{16}$. This is done as follows:

```
16)13.0000(.8125
   12 8
   ----
     20
     16
     --
      40
      32
      --
       80
       80
       --
```

In this case, the quotient is exact and the decimal equivalent of $\frac{13}{16}$ is .8125, that is, $\frac{13}{16}$ = .8125, "8,125 ten thousandths."

As another example, we will find the decimal equivalent of the common fraction $\frac{113}{330}$. Here we have:

```
330)113.0000000(.3424242
     99 0
     ----
     14 00
     13 20
     -----
        800
        660
        ---
       1400
       1320
       ----
          800
          660
          ---
         1400
         1320
         ----
            800
            660
            ---
```

The fraction $\frac{113}{330}$ has no exact decimal equivalent. No matter how many decimal places it may be carried out there will always be a remainder. However, we notice that the decimal obtained in our division has the cycle of repeating digits 42, 42, etc. after the 3. In general, when a fraction is converted to a decimal we have two possibilities:

a) the decimal equivalent is a terminating decimal.

b) the decimal equivalent is a repeating decimal.

It must be understood that, in the case of a repeating decimal, the decimal representation does not represent the exact value of the fraction. This may be indicated in the following ways:

$$\frac{113}{330} = .3424242\ldots\ldots$$

The dots indicate that this process may be continued without end

$$\frac{113}{330} \doteq .3424242$$

The equality sign with the dot over it meaning "approximately equals."

Consideration of this fraction introduces the inverse process to the one discussed above: to convert a decimal fraction into a common fraction. In this chapter, we shall consider only one type of decimal fraction, the terminating decimal fraction. For example,

$$.625 = \frac{625}{1000} = \frac{5}{8};\ .8125 = \frac{8125}{10000} = \frac{13}{16};\ .75 = \frac{75}{100} = \frac{3}{4}.$$

From these examples we get the following rule.

RULE II: *To convert an exact terminating decimal fraction into a common fraction, write the given decimal as a numerator, omitting the decimal point, and place under it the denominator* 10, 100, 1000, *etc., indicated by the number of decimal places in the given decimal fraction, and reduce the resulting common fraction to its lowest terms.*

A final illustration of this rule is: $.0048 = \frac{48}{10000} = \frac{12}{2500} = \frac{3}{625}$. Here the figures used as numerator are 48, and as the decimal fraction contains four decimal places the denominator is 10000.

35. Exercises.

1. Reduce to lowest terms: $\frac{72}{120}$; $\frac{84}{96}$; $\frac{168}{252}$; $\frac{75}{135}$; $\frac{156}{208}$.
2. Reduce to mixed numbers: $\frac{154}{9}$; $\frac{349}{13}$; $\frac{588}{24}$; $\frac{978}{56}$; $\frac{3828}{836}$.
3. Express as improper fractions: $15\frac{4}{5}$; $15\frac{7}{8}$; $24\frac{3}{4}$; $20\frac{3}{7}$; $8\frac{5}{12}$.
4. Express with common denominator: $\frac{3}{4}$, $\frac{5}{6}$ and $\frac{7}{8}$; $\frac{4}{5}$, $\frac{5}{6}$ and $\frac{9}{10}$; $\frac{6}{14}$, $\frac{10}{24}$ and $\frac{13}{42}$.

Add the following fractions:

5. $\frac{7}{12}$, $\frac{4}{12}$, $\frac{5}{12}$ and $\frac{11}{12}$.
6. $\frac{2}{3}$, $\frac{3}{4}$, $\frac{5}{6}$ and $\frac{7}{8}$.
7. $41\frac{1}{2}$, $105\frac{2}{9}$, $300\frac{3}{4}$, $241\frac{3}{5}$ and $472\frac{1}{5}$.

Subtract the following:

8. $\frac{21}{56}$ from $\frac{35}{56}$.
9. $\frac{3}{8}$ from $\frac{5}{6}$.
10. $\frac{19}{65}$ from $\frac{14}{39}$.
11. $3\frac{9}{14}$ from $28\frac{16}{63}$.
12. $4\frac{3}{7}$ from 75.

Multiply the following:

13. $8 \times \frac{3}{9}$; $\frac{2}{75} \times 15$; $8 \times \frac{3}{4}$.
14. $\frac{2}{3} \times \frac{5}{7}$; $\frac{9}{10} \times \frac{7}{8}$; $\frac{4}{5} \times \frac{3}{4}$.
15. $\frac{2}{3} \times \frac{5}{6} \times \frac{9}{15} \times \frac{7}{8}$.
16. $\frac{2}{3} \times \frac{7}{8} \times \frac{15}{28} \times \frac{4}{11} \times \frac{44}{75}$.
17. $2\frac{2}{5} \times 2\frac{4}{7} \times \frac{2}{11} \times \frac{5}{108} \times 1\frac{7}{15} \times 26\frac{1}{4}$.

Perform the following divisions:

18. $\frac{12}{35} \div 4$; $\frac{10}{11} \div 5$; $\frac{15}{17} \div 3$.
19. $10 \div \frac{2}{7}$; $4 \div \frac{1}{3}$; $16 \div \frac{4}{5}$.
20. $\frac{26}{27} \div \frac{8}{9}$; $\frac{12}{55} \div \frac{9}{77}$; $\frac{15}{24} \div \frac{5}{6}$.
21. $56 \div 1\frac{5}{9}$; $1\frac{7}{8} \div 1\frac{1}{8}$; $1\frac{14}{91} \div \frac{35}{52}$.

Find, by cancellation, the values of the following expressions:

22. $$\frac{18 \times 6 \times 4 \times 42}{4 \times 9 \times 3 \times 7 \times 6}.$$

23. $$\left(\frac{21 \times 8 \times 60}{7 \times 12 \times 3}\right) \times \left(\frac{8 \times 6}{8 \times 3}\right).$$

24. $$\left(\frac{16 \times 5 \times 14 \times 40}{40 \times 24 \times 50}\right) \div \left(\frac{20 \times 7 \times 10}{16 \times 60 \times 50}\right).$$

25. Convert to decimal fractions: $\frac{7}{12}$; $\frac{15}{16}$; $\frac{12}{25}$; $\frac{53}{64}$.
26. Convert to common fractions: 0.85, .375, .3175, .864, .6875.

Chapter 7

POWERS AND ROOTS

36. Powers and Exponents. If a certain number is the product of two or more other numbers, those other numbers are *factors* of the product number, as we have seen. If the several factors of a product number are *equal* that number is said to be a *power* of the repeated factor. Thus, $5 \times 5 = 25$, and 25 is a power of 5; $2 \times 2 \times 2 = 8$, and 8 is a power of 2. The power is designated by stating the number of times the factor is used. In $5 \times 5 = 25$, 5 is used twice as a factor and 25 is said to be the *second power* of 5; $2 \times 2 \times 2 = 8$, and 8 is the *third power* of 2. The 5 is said to be *raised* to the second power, and similarly 2 is raised to the third power. Similarly, to raise 3 to the fourth power, 3 is used as a factor four times: $3 \times 3 \times 3 \times 3 = 81$, and 81 is the fourth power of 3. The operation of raising a number to a power is called *involution*.

In order to avoid writing out the multiplication in full, as

$$5 \times 5$$
$$2 \times 2 \times 2$$
$$3 \times 3 \times 3 \times 3$$

or

$$5 \times 5 \times 5 \times 5 \times 5,$$

these are indicated by writing the number with a smaller figure just above and to the right to show the number of times the factor is used. Thus,

5×5 is written 5^2
$2 \times 2 \times 2$ " " 2^3
$3 \times 3 \times 3 \times 3$ " " 3^4
$5 \times 5 \times 5 \times 5 \times 5$ " " 5^5, etc.

These are read as follows:

5^2 is "5 to the second power,"
or "5 to the second,"
2^3 is "2 to the third,"
3^4 is "3 to the fourth,"
5^5 is "5 to the fifth," etc.

Using this form of notation, instead of writing $5 \times 5 = 25$, $2 \times 2 \times 2 = 8$, etc., we can write

$$5^2 = 25,$$
$$2^3 = 8,$$
$$3^4 = 81,$$
$$5^6 = 15625,$$
$$7^3 = 343,$$
$$8^2 = 64, \text{ etc.}$$

The small figure used to indicate the power to which a number is raised is called an *exponent*, and the number which is raised to that power is called the base number, or simply the *base*. Thus, in 5^2 the 2 is the exponent of 5 which is the base.

Exponents must not be confused with factors. Thus 5^2 does NOT mean $5 \times 2 = 10$ but $5 \times 5 = 25$, not 5 times 2 but 5 used *two times* as a factor; 7^3 does NOT mean $7 \times 3 = 21$, but $7 \times 7 \times 7 = 343$, 7 used *three times* in multiplication.

Now the reader may recall from elementary arithmetic the fact (which we shall find again later) that the area of a square figure is found by multiplying the length of the side by itself. Thus, if a square is 5 inches on a side the area is $5 \times 5 = 25$ square inches. That is, the area of a square is equal to the second power of the number representing its side. For this reason, the second power of a number is called the *square* of the number, and when a number is raised to the second power it is said to be *squared*. Thus, in the operation $5^2 = 25$ the 5 is squared and we say that "the square of 5 is 25" or that "5 squared is 25."

Similarly, the volume of a cube (square block) is found by using three times as a factor the number representing the length of its edge. Thus, a square block 2 inches on edge has a volume of $2 \times 2 \times 2 = 8$ cubic inches. That is, the volume of a cube is the third power of the number representing its edge. For this reason, the third power of a number is called the *cube* of the number, and the number raised to the third power is said to be *cubed*. Thus, in the operation $2^3 = 8$ the 2 is cubed and we say that "the cube of 2 is 8" or that "2 cubed is 8."

There are no similar or corresponding names for other powers.

Any number may be raised to any power by repeated multiplication. Thus 958^3 is found by carrying out the multiplication $958 \times 958 \times 958$ to be equal to 879,217,912; similarly $9.1^4 = 6{,}857.4961$; etc.

37. Higher Powers. From the definition of a power, as just seen, a number may be raised to any power by repeated multiplication. For higher powers, however, such an operation is long and tedious, but may be shortened in certain cases. Consider the fourth power of 3. Now

$$3^4 = 3 \times 3 \times 3 \times 3 = (3 \times 3) \times (3 \times 3), \text{ and } 3 \times 3 = 3^2.$$

Therefore, $3^4 = (3^2) \times (3^2)$, and this is itself the square of (3^2). That is, *the fourth power of a number is the square of the square* of the number. In order to raise any number to the fourth power, therefore, it is only necessary to square the number, and then square this result; thus only two multiplications are required instead of three.

Similarly, $2^6 = 2 \times 2 \times 2 \times 2 \times 2 \times 2 = (2 \times 2 \times 2) \times (2 \times 2 \times 2) = (2^3) \times (2^3) = 8 \times 8 = 64$, and so for any other number. That is, *the sixth power of any number is the square of the cube* of the number. By utilizing this principle, a number can be raised to the sixth power by four multiplications, instead of the five required by straightforward multiplication.

In the same way it is easy to see that

The eighth power of a number is the square of the fourth power.
The ninth power is the cube of the cube.
The twelfth power is the cube of the fourth power, or *the square of the sixth.*

In this manner the operation of finding many higher powers of a number can be shortened by combining certain of the lower powers whose exponents are *factors* of the exponent of the higher power.

The fifth power of a number, say 3, is $3^5 = 3 \times 3 \times 3 \times 3 \times 3 = (3 \times 3 \times 3 \times 3) \times 3 = 3^4 \times 3$, that is, *the fourth power of the number multiplied by the number itself.* The *tenth power is* then *the square of the fifth* power. Similarly, the seventh power is the product of the sixth power by the number itself, or the product of the fourth power by the cube.

Other such combinations will suggest themselves to the student who finds it necessary to raise numbers to higher powers. It will be observed that many of these combinations involve the square and the cube. For this reason, tables have been prepared which give the squares and cubes of whole numbers, and by placing the decimal points according to the usual rules of multiplication, such a table will serve for decimals.

A table of the squares and cubes of all numbers from 1 to 1560 is given at the end of this book, Table III. For any number in the first (left hand) column, and on the same line with the number, the square or cube is found in the column headed *squares* or *cubes* respectively. Thus, the square of 1175 is 1380625; the cube of 397 is 62570773.

38. Roots and Root Indexes. When a number is the product of several equal factors (a power of that repeated factor) each of the equal factors, or the repeated factor, is called a *root* of the number. Thus, $25 = 5 \times 5$ and 5 is a root of 25; $8 = 2 \times 2 \times 2$ and 2 is a root of 8; etc. Stated in other words, a root of any given number is that number which must be raised to a certain power to produce the given number. The root is designated by naming the exponent of the power to which it must be raised to produce the given number.

Thus, 5×5 or $25 = 5^2$ and 5 is called the second root of 25; $8 = 2^3$ and 2 is the third root of 8; $1296 = 6^4$, and 6 is the fourth root of 1296; etc.

In analogy with the second and third powers, the second root of a number is called the *square root* and the third root is called the *cube root.*

A special form is used in writing to indicate roots: the third or cube root of 64, for example, is written $\sqrt[3]{64}$; the fifth root of 32 is written $\sqrt[5]{32}$; etc. The symbol $\sqrt{\ }$ is called the root or *radical* sign, and the small figure placed in the "hook" of the radical sign to indicate the number of the root, is called the *root index.* Thus in the indicated fifth root of 32, $\sqrt[5]{32}$, the root index is 5.

Since any number is its own first power and its own first root it is not necessary to use an exponent or a radical sign and root index for the first power or root. Thus $9^1 = 9$ and $\sqrt[1]{9} = 9$.

Then, since there is no other root with a whole number index less than 2 to be distinguished, it is not necessary to show the index of the second or square root. Therefore, the square root is written without the index. Thus $\sqrt{9}$ is the square root of 9. In the case of the third or cube root, the fourth root, and all others, it is necessary to show the index with the radical sign. Thus: $\sqrt[3]{27}$, $\sqrt[4]{16}$, $\sqrt[5]{243}$, etc.

When a root of a number is found or calculated by any means, the root is said to be *taken* or *extracted.* The general operation of root extraction is called *evolution.*

While any power of any number may be found simply by repeated multiplication, there is no such direct method of extracting roots in general. There are rules for extracting the square and cube roots but even these are somewhat long and tedious; there are no such direct rules for higher roots, however. The rules for square and cube roots are given and their use illustrated in the following articles.

39. Square Root. The principle and method of extracting square roots will be explained in some detail on account of the importance of the operation and the frequency of its occurrence.

The squares of the numbers

$$1,\ 2,\ 3,\ 4,\ 5,\ 6,\ 7,\ 8,\ 9$$

are

$$1,\ 4,\ 9,\ 16,\ 25,\ 36,\ 49,\ 64,\ 81.$$

Therefore, the square *root* of a number of one or two figures is a number of one figure.

The squares of the numbers

$$10,\quad 11,\quad 12,\ \ldots\ldots,\quad 97,\quad 98,\quad 99$$

are

$$100,\ 121,\ 144,\ \ldots\ldots,\ 9409,\ 9604,\ 9801.$$

Therefore, the square *root* of a number of three or four figures is a number of two figures.

Similarly, the square root of a number of five or six figures is a number of three figures, and so on.

We have, therefore, the following important result, or

RULE: *If the figures of a whole number are marked off in groups of two figures each, beginning at the right, any single figure remaining at the left being counted as a group, the number of figures in the square root is the same as the number of groups.*

Thus, marked off in groups, 18190225 becomes 18,19,02,25; therefore, $\sqrt{18190225}$ contains four figures. It is, in fact, 4265. Similarly, $\sqrt{3{,}47{,}07{,}69}$ contains four figures; it is 1863.

Let us consider the square of the number 43; this is $43 \times 43 = 1849$. Therefore, $\sqrt{1849} = 43$. Now $43 = 40 + 3$ and, therefore, 43×43 is the same as $(40 + 3) \times (40 + 3)$. Multiplying each term or number in the second parenthesis by each term in the first this product is expressed as follows:

$$\begin{array}{r} 40 + 3 \\ 40 + 3 \\ \hline 120 + 9 \\ 1600 + 120 \\ \hline 1600 + 240 + 9 \end{array} \qquad\qquad \begin{array}{r} 43 \\ 43 \\ \hline 129 \\ 172 \\ \hline 1849 \end{array}$$

and $1600 + 240 + 9 = 1849$ is the same as $(40)^2 + (2 \times 40 \times 3) + (3)^2$. Similarly,

$$(22)^2 = (20 + 2) \times (20 + 2) = 20^2 + 2 \times 20 \times 2 + 2^2;$$
$$(31)^2 = (30 + 1) \times (30 + 1) = 30^2 + 2 \times 30 \times 1 + 1^2; \text{ etc.}$$

That is, the square of the sum of any two numbers equals the sum of their separate squares plus twice their product.

Now to reverse the process, obtaining

$$\sqrt{1849} = \sqrt{(40)^2 + (2 \times 40 \times 3) + (3)^2} = 40 + 3 = 43,$$

the operation is carried out as follows, the explanation following:

$$\begin{array}{r l}
 & (40)^2 + 2 \times 40 \times 3 + (3)^2 \underline{(40 + 3} \\
 & (40)^2 \\
\hline
(2 \times 40) + 3) & 2 \times 40 \times 3 + (3)^2 \\
 & 2 \times 40 \times 3 + (3)^2 \\
\hline
\end{array}$$

Here we notice that 40 is the square root of the first part, $(40)^2$, and write down the 40 at the right. This 40 is squared and the square subtracted from the $(40)^2 + (2 \times 40 \times 3) + (3)^2$, leaving $(2 \times 40 \times 3) + (3)^2$. The 40 is then doubled and the 2×40 is used as a *trial divisor*, the partial dividend being the remainder $(2 \times 40 \times 3) + (3)^2$ as in long division. The quotient is 3. This quotient is written as the second part of the result and is also added to the trial divisor, giving $(2 \times 40) + 3$. This complete divisor is contained in the partial dividend 3 times and is multiplied by this 3, giving $3 \times (2 \times 40) + 3 \times 3$, or $(2 \times 40 \times 3) + (3)^2$. This product is written under the dividend and subtracted from it. The remainder is zero, the operation is complete, and the result is $40 + 3$. That is, $\sqrt{(40)^2 + (2 \times 40 \times 3) + (3)^2} = 40 + 3$, or $\sqrt{1849} = 43$.

Omitting the zeros and plus signs, that is, writing the numbers all together in the usual manner instead of showing the separate steps in parts, the complete operation is equivalent to the following: The number is divided into two groups, 18, 49. The greatest exact square in the first group, 18, is 16 and its square root is 4. This 4 is written as the first figure of the desired result, is then squared and the square written down under the first group and subtracted. With the remainder is written down the next group, 49, giving a dividend of 249.

$$\begin{array}{r l}
 & 18,49(43 \\
 & \underline{16} \\
83) & 2\ 49 \\
) & \underline{2\ 49} \\
\end{array}$$

The part of the root already found, 4, is doubled to form a partial divisor and this partial divisor, 8, is divided into the first part of the dividend 249. The quotient 3 is written down as the next figure of the desired root. The 3 is also written beside the partial divisor (at the left) to give the complete divisor 83. This complete divisor is then multiplied by the last figure of the root, 3, and the product written under the dividend and subtracted from it. The remainder is zero, the operation is complete, and the result is 43. That is, $\sqrt{1849} = 43$.

If there were another group of figures in the original number it would next be drawn down and the process repeated, now doubling the 43 to form the partial divisor. This is illustrated without explanation in the following example:

To find: $\sqrt{59536}$

Result: $\sqrt{59536} = 244$

```
      5,95,36(244
      4
      -------
   44)1 95
     )1 76
      -------
  484)19 36
      )19 36
      -------
```

From the preceding examples and explanations we can now write the following:

RULE FOR SQUARE ROOT: (i) *Mark off the number whose square root is to be found, in groups of two figures, beginning at the right.*

(ii) *Determine by inspection the greatest exact square in the left-hand group and write its square root as the first figure of the required root.*

(iii) *Subtract the square of this figure from the left-hand group and with the remainder draw down the next group to form a partial dividend.*

(iv) *Double that part of the root already found for a partial divisor and divide it into the first part of the dividend; the quotient* (*or the quotient diminished*) *is the next figure of the required root.*

(v) *Write this last figure beside the partial divisor* (*on the left*) *to form a complete divisor, and multiply this complete divisor by this last figure, writing the product under the dividend.*

(vi) *Subtract this product from the dividend, draw down the next group to form a new dividend, double that part of the root already found to form a partial divisor, and proceed as before.*

(vii) *If there is no remainder from the dividend formed with the last group in the original number, the operation is ended and the root is exact. If there is a remainder write a decimal point after that part of the root already found, draw down two ciphers as the next group, proceed as before and continue as long as desired.*

(viii) *If at any time the partial divisor is greater than the first part of the dividend, write a zero as the corresponding figure of the root, draw down the next group of figures and proceed as before.*

(ix) *If the given number is a decimal number begin at the decimal point and mark off groups of two figures in both directions; if the last group on the right contains only one figure annex a zero. Then proceed as before, placing a decimal point in the root when that in the number is reached.*

One more illustration is added, showing the application of the complete rule. To find $\sqrt{526930.81}$

```
                52,69,30.81(725.9
                49
              -----------
            142) 3 69
               ) 2 84
                 ---------
            1445)85 30
                )72 25
                 ---------
           14509)13 05 81
                )13 05 81
                 --------
```

The result is, therefore, $\sqrt{526930.81} = 725.9$.

40. Cube Root. The rule for cube root will be stated directly, without preliminary analysis, and its use will then be illustrated. This procedure is followed because the derivation of the rule is somewhat long and involved and also because it is not so much used as is the rule for the square root.

RULE FOR CUBE ROOT: (i) *Mark off the number whose root is to be found in groups of three figures, beginning at the decimal point and proceeding both ways, and adding ciphers to the decimal part when the last group does not contain three figures.*

(ii) *Find the greatest cube in the left-hand group and write its cube root as the first figure in the required root.*

(iii) *Subtract the cube of this figure from the left-hand group and with the remainder draw down the next group to form a partial dividend.*

(iv) *At the left of this dividend write three hundred times the square of the first figure of the root for a trial divisor; divide the dividend by this trial divisor, and write the quotient for a trial figure in the root.*

(v) *For the complete divisor add to the partial divisor the following two numbers: thirty times the product of the first figure and the trial figure, and the square of the trial figure.*

(vi) *Multiply the complete divisor by the trial figure; if the product is less than the dividend subtract it from the dividend, and with the remainder bring down the next group for a new dividend.*

(vii) *If, however, the last product is greater than the dividend, replace the trial figure in the root by the next smaller, form a new dividend as in steps* (iv) *and* (v), *and find a new remainder to form the new dividend in* (vi).

(viii) *Using the two figures of the root now as the "first figure" repeat step* (iv) *to find the next trial figure.*

(ix) *Repeat steps* (v) *and* (vi), *and* (vii) *if necessary, and continue until there are no more groups in the original number, or as long as desired by annexing new groups of zeros.*

(x) *Point off to the left of the decimal point in the root as many figures as there are groups to the left of the point in the original number.*

As an example of the use of this rule we find here the cube root of the number 78,347,809.639.

		78,347,809.639(427.9
		64
$300 \times 4^2 =$	4800	14 347
$30 \times 4 \times 2 =$	240	
$2^2 =$	4	
	5044	10 088
$300 \times 42^2 =$	529200	4 259 809
$30 \times 42 \times 7 =$	8820	
$7^2 =$	49	
	538069	3 766 483
$300 \times 427^2 =$	54698700	493 326 639
$30 \times 427 \times 9 =$	115290	
$9^2 =$	81	
	54814071	493 326 639

The result is $\sqrt[3]{78,347,809.639} = 427.9$.

41. Higher Roots. It was stated in article 38 that there are no simple direct rules for extracting roots other than the second and third, or square and cube, roots. We have seen in article 37, however, that many higher powers of numbers may be found by repeated squaring or cubing of the lower roots. Since evolution is the inverse of involution it would therefore seem that many higher roots could be found by repeated extractions of square and cube roots. Such is indeed the case.

Thus since the fourth *power* is the square of the square,

The fourth root is the square root of the square root.

Similarly,

The sixth root is the cube root of the square root;

The eighth root is the square root of the fourth root, or

The eighth root is the square root of the square root of the square root;

The ninth root is the cube root of the cube root; etc.

In general with the rules for square root and cube root, any higher root can be extracted *if the root index is a power of* 2 *or* 3 *or a multiple of both* 2 *and* 3. Thus, the roots of index 2, 3, 4, 6, 8, 9, 12, 16, 18, etc., can be found by repeated extractions of square and cube roots, but not the roots of index 5, 7, 11, 13, 15, 17, 19, 20, etc.

Since the square and cube roots are of such importance tables have been calculated which give directly the square and cube roots of whole numbers, and by placing the decimal point in any case in accordance with the usual rules, such a table will serve for decimal numbers also. Table III, in the back of this book, gives the square and cube roots of all numbers from 1 to 1560. These are found on the line with the number itself in the first column and in the column headed *square roots* or *cube roots*, respectively.

By taking numbers in the *squares* or *cubes* columns in this table and reversing the readings, the square and cube *roots*, respectively, are found in the first column of the table.

42. Remarks on Powers and Roots. We have seen that *any* power of any number can be found by repeated multiplication, and how certain higher powers of a number can be found by combining certain of its lower powers. Thus involution is always possible, even though the process becomes long and tedious for large numbers or higher powers.

We have seen also that by means of the rules for square and cube roots many other higher roots may be calculated. Not all roots can be found by such methods, however, and even those which can be found require much time and labor, since the applications of the rules for square and cube roots are themselves long and tedious operations.

There is a method, however, by means of which *any* root *or* power of any number may be very simply and easily found. The full theory and exact use of this method are somewhat involved but in the next two chapters are given sufficient description for almost any ordinary needs in calculation. This method is not only useful in finding powers and roots but can also be used to shorten many long multiplications and divisions.

43. Exercises.

Raise the following numbers to the powers indicated, using the method stated:

1. 859^2; 51^3; 12^7; by straightforward multiplication.
2. 21^4; 23^6; 9^8; by combination of lower powers.

Find the following indicated roots:

3. $\sqrt{101761}$; $\sqrt{1522756}$; $\sqrt{704.3716}$; $\sqrt{4012}$.
4. $\sqrt[3]{941192}$; $\sqrt[3]{1860867}$; $\sqrt[3]{6751.269}$; $\sqrt[3]{4012}$.
5. $\sqrt[4]{31640625}$; $\sqrt[6]{387420489}$; $\sqrt[9]{40353607}$.

(A simpler method of finding roots and powers, and one which may be used when the methods of articles 39–41 do not apply, is to be explained in a later chapter. More exercises will then be given for practice.)

Chapter 8

LOGARITHMS

44. Use of Exponents in Calculation. By definition of a power, $2^2 = 2 \times 2 = 4$, $2^3 = 2 \times 2 \times 2 = 2 \times 4 = 8$, $2^4 = 2 \times 2 \times 2 \times 2 = 2 \times 2^3 = 2 \times 8 = 16$, $2^5 = 2 \times 2^4 = 2 \times 16 = 32$, $2^6 = 2 \times 32 = 64$, $2^7 = 2 \times 64 = 128$, etc., each successive power of 2 being obtained by doubling the preceding one. Similarly, $3^2 = 9$, $3^3 = 3 \times 9 = 27$, $3^4 = 3 \times 27 = 81$, $3^5 = 3 \times 81 = 243$, etc., each successive power of 3 being obtained by multiplying the preceding one by 3. In the same way the successive powers of 10 or any number can be found. In this way a list or table of the successive powers can be made out for any number. The table at the right gives all the successive powers of 2 up to the 20th. By means of such a table many simple calculations can be very easily performed by noting certain characteristic properties of the series of numbers in the table.

$$\begin{aligned}
2^0 &= 1\\
2^1 &= 2\\
2^2 &= 4\\
2^3 &= 8\\
2^4 &= 16\\
2^5 &= 32\\
2^6 &= 64\\
2^7 &= 128\\
2^8 &= 256\\
2^9 &= 512\\
2^{10} &= 1024\\
2^{11} &= 2048\\
2^{12} &= 4096\\
2^{13} &= 8192\\
2^{14} &= 16384\\
2^{15} &= 32768\\
2^{16} &= 65536\\
2^{17} &= 131072\\
2^{18} &= 262144\\
2^{19} &= 524288\\
2^{20} &= 1048576
\end{aligned}$$

Thus let us multiply 4096×64. By carrying out the multiplication in the usual way the product is found to be 262144. It is to be noted, however, that 64 is the 6th power of 2, and 4096 is the 12th power of 2, while the product is the 18th power of 2, that is, 2^{18}. Similarly, if we multiply $2^8 = 256$ by $2^9 = 512$ the product is $2^{17} = 131072$; that is, $2^{12} \times 2^6 = 2^{18}$ and $2^8 \times 2^9 = 2^{17}$. In general, if any power of 2 is multiplied by any other power of 2, the product is a power of 2 whose exponent is the sum of the exponents corresponding to the multiplicand and the multiplier.

In order to multiply any number in the list by itself, the same exponent is used twice. Thus, 128×128 is the same as $2^7 \times 2^7 = 2^{14} = 16384$. That is, in order to square a number, the exponent corresponding to it is doubled. Similarly, in order to raise a number to the third or fourth power, the exponent corresponding to it is multiplied by 3 or 4, and so on. In general, to raise a number to any power, the exponent corresponding to that number is multiplied by the exponent of the stated power. Reversing the procedure of multiplication by *adding* exponents, it is seen at once that division is performed by *subtracting* exponents. Similarly, reversing the procedure of involution (multiplying exponents), the extraction of a root is performed by dividing the exponent corresponding to the number by the index of the root. Thus, to find the 5th root of 1048576 the exponent corresponding to 1048576, which is 20, is divided by the root index 5, $20 \div 5 = 4$. The number corresponding to the resulting exponent 4 is the desired root; it is 16. That is $\sqrt[5]{1048576} = 16$.

45. Logarithms. When it is desired to raise a certain number to a specified power, as 5^2, 10^2, 10^3, 2^3, 2^{10}, the operation is thought of as involving the base number, and the exponent indicates the number of times the base is used as a factor. Thus, 5^2 means 5×5, 2^3 means $2 \times 2 \times 2$, 10^3 means $10 \times 10 \times 10$, 2^{10} means $2 \times 2 \times 2$ and so on 10 times.

When the *result* of the operation is considered, however, the relation of the exponent to *this* number is viewed differently. Thus, $5^2 = 25$ and the 2 is the indicator of a certain operation which produces 25; $10^3 = 1000$ and the 3 is related to the 1000 in the same way; $2^{10} = 1024$ (from the table) and again it is the relation of the 10 to the 1024 which is of importance, rather than the relation of the 10 to the 2. This relation is indicated by giving the exponent a new name.

When, in the operation $2^{10} = 1024$, it is desired to express the relation of the 10 to the 1024 it is said that 10 is the *logarithm* of 1024 to the base 2. Thus, when we write $2^{10} = 1024$, 10 is the exponent of the 2 but it is the logarithm of 1024. This is abbreviated by writing $10 = \log 1024$, or, if it is desired also to indicate the base number, $\log_2 1024$. Thus:

$$2^{10} = 1024, \qquad 10 = \log_2 1024,$$

the base being indicated by the small figure just below and to the right of the abbreviation "log." Similarly, since $5^2 = 25$, $2 = \log_5 25$, or as it is more generally stated, $\log_5 25 = 2$; $3^4 = 81$, or $81 = 3^4$, and $\log_3 81 = 4$; $1000 = 10^3$ and $\log_{10} 1000 = 3$; etc.

Number	Log.
1	0
2	1
4	2
8	3
16	4
32	5
64	6
128	7
256	8
512	9
1024	10
2048	11
4096	12
8192	13
16384	14
32768	15
65536	16
131072	17
262144	18
524288	19
1048576	20

Since any positive number may be used as a base number to produce any required number as a power, the base is not so important except that a convenient one should be chosen. The important consideration is the logarithm, originally called the exponent of the base. Thus, in the operations of article 44, the numbers in the right-hand column which were multiplied, divided, raised to powers, had roots extracted; these are ordinary *numbers*, and it was the exponents of the base 2, or the *logarithms of the numbers*, which so greatly simplified the operations. If it is understood that a certain base number is used, then it is not necessary to write it each time and the table may be arranged to show simply the numbers and their logarithms. In this way the table of article 44, given in terms of the exponents of the base, is given above at the right as a table of numbers and their logarithms referred to the base 2. From this table we can read at a glance:

$$\log_2 1024 = 10$$
$$\log_2 131072 = 17$$
$$\log_2 8 = 3$$
$$\log_2 16 = 4$$
$$\log_2 8192 = 13, \text{ etc.}$$

The operations performed in article 44 are now seen to be: for

multiplication,	addition of logs.
division,	subtraction of logs.
powers,	multiplication of logs.
roots,	division of logs.

Example: Multiply 512 by 2048

$$\begin{aligned} \log_2 512 &= 9 \\ \log_2 2048 &= 11 \\ \hline \text{log of Product} &= 20 \\ \text{Product} = 2^{20} &= 1048576 \end{aligned}$$

Example: Divide 65536 by 1024

$$\begin{aligned} \log_2 65536 &= 16 \\ \log_2 1024 &= 10 \\ \hline \text{log of Quotient} &= 6 \\ \text{Quotient} &= 2^6 = 64 \end{aligned}$$

Example: Find the Value of 64^3

$$\begin{aligned} \log_2 64 &= 6 \\ \log_2 64^3 &= 3 \times 6 = 18 \\ \text{log of power} &= 18 \\ \text{Hence } 64^3 &= 2^{18} = 262144 \end{aligned}$$

Example: Find the Value of $\sqrt[5]{32768}$

$$\begin{aligned} \log_2 32768 &= 15 \\ \log_2 32768^{\frac{1}{5}} &= \tfrac{1}{5}(15) = 3 \\ \text{log of root} &= 3 \\ \text{Hence } \sqrt[5]{32768} &= 2^3 = 8 \end{aligned}$$

46. Common Logarithms. As already seen, any positive number other than 1 may be used as a base for logarithms. If, however, the base is small, the logarithm of a specified number may be large. Thus, $\log_2 1{,}048{,}576 = 20$ and for larger numbers the logarithm may be much greater, the table in the preceding article being, of course, only a small and incomplete table. If we use a larger number as base, the logarithm of any specified number will be smaller. Thus, if 10 is taken as base $\log_{10} 1{,}000{,}000 = 6$ because $10^6 = 1{,}000{,}000$, and $\log_{10} 10{,}000{,}000 = 7$. In actual practice, the number 10 is always used as base and logarithms to base 10 are called *common logarithms.* On account of the importance of the number 10 in connection with decimal numbers, the use of 10 as base possesses many advantages not obtainable in ordinary calculation with other bases, as will appear later.

When 10 is used as base, it is not necessary to indicate it in writing logarithms, it being understood that 10 is the common base. Thus, instead of $\log_{10} 1000 = 3$, it is sufficient to write $\log 1000 = 3$, and similarly for the common logarithm of any number. In what follows, this practice will be followed.

From the definition of logarithms as found from exponents it is easy to prepare the following table of common logarithms:

$10^0 = 1$	$\log 1 = 0$
$10^1 = 10$	$\log 10 = 1$
$10^2 = 100$	$\log 100 = 2$
$10^3 = 1000$	$\log 1000 = 3$
$10^4 = 10000$	$\log 10000 = 4$
$10^5 = 100000$	$\log 100000 = 5$
etc.	etc.

Now the decimal fraction .1 is by definition $\frac{1}{10}$, and .01 is $\frac{1}{100}$; similarly $.001 = \frac{1}{1000}$; $.0001 = \frac{1}{10000}$; etc. In accordance with the method of writing decimal fractions given in article 3, and in comparison with $10 = 10^1$, $100 = 10^2$, $1000 = 10^3$, etc., these fractions are written as $.1 = \frac{1}{10^1} = 10^{-1}$; $.01 = \frac{1}{10^2} = 10^{-2}$; $.001 = \frac{1}{10^3} = 10^{-3}$; etc. Using the logarithmic form of expressing exponents, therefore, we can prepare a table of logarithms of decimal fractions as follows:

$10^{-1} = .1$	$\log .1 = -1$
$10^{-2} = .01$	$\log .01 = -2$
$10^{-3} = .001$	$\log .001 = -3$
$10^{-4} = .0001$	$\log .0001 = -4$
$10^{-5} = .00001$	$\log .00001 = -5$
etc.	etc.

Logarithms written in this form are called *negative* logarithms to distinguish them from the logarithms of whole numbers, which are called *positive* logarithms. For convenience in using logarithms -1, -2, -3, etc. are usually written as $9.0000 - 10$, $8.0000 - 10$, $7.0000 - 10$, respectively. Thus, numbers greater than 1 have positive logarithms while decimal fractions have negative logarithms:

$$\log .1 = 9.0000 - 10; \quad \log .01 = 8.0000 - 10, \quad \text{etc.}$$

Now, the tables of common logarithms given above are far from complete as they only show the logarithms of the even tens, hundreds, thousands, etc., and the even tenths, hundredths, thousandths, etc. The numbers between 1 and 10, between 10 and 100, 100 and 1000, or between .1 and .01, .01 and .001, etc., to complete the table will, however, not have whole number logarithms, as those inserted numbers are not exact powers of 10, either positive or negative. Thus, since log 100 = 2 and log 1000 = 3 the logarithms of numbers between 100 and 1000 will be between 2 and 3, and each will be 2 plus a *fraction.* Also, since log .01 = −2 and log .001 = −3 the logarithms of decimal fractions between .01 and .001 will be between −2 and −3 and each will be −2 minus a fraction or −3 plus a fraction. The whole number part of such a logarithm is called the *characteristic* of the logarithm and the fractional part (expressed as a decimal) is called the *mantissa* of the logarithm. For example, the number 2473 lies between 1,000 and 10,000 and therefore its logarithm lies between 3 and 4; it has been found by special methods to be 3.3932 (to four decimal places). We write, therefore, log 2473 = 3.3932, and the characteristic of log 2473 is 3 while the mantissa of log 2473 is .3932.

The statement that log 2473 = 3.3932 means that

	$10^{3.3932} = 2473$,	written	$\log 2473 = 3.3932$.
Similarly,	$10^{2.3932} = 247.3$,	hence	$\log 247.3 = 2.3932$
	$10^{1.3932} = 24.73$		$\log 24.73 = 1.3932$
	$10^{0.3932} = 2.473$		$\log 2.473 = 0.3932$
	$10^{9.3932-10} = .2473$		$\log .2473 = 9.3932 - 10$
	$10^{8.3932-10} = .02473$		$\log .02473 = 8.3932 - 10$
	etc.		etc.

From the first tables of common logs of even tens, hundreds, tenths, hundredths, etc., and the table just given showing the logarithms of other numbers, it is seen at once that the characteristic of the logarithm of a specified number is determined by noting whether the number is between 1 and 10, between 10 and 100, 100 and 1000, etc., or between 1 and .1, .1 and .01, .01 and .001, etc., that is, by noting *how many figures there are to the left of the decimal point in the number*, or *how many zeros there are to the right of the decimal point in a decimal fraction.* The *mantissa* of the logarithm, however, depends only *on the sequence of figures in the number* and not on the position of the decimal point.

Thus, the mantissa is the same for log 24730, log 2473, log 24.73, log .002473, etc., and is .3932. The characteristic is, however, for log 24730, 4; for log 2473, 3; for log 24.73, 1; for log .002473, 7 − 10 (or −3), etc. In making up a table of common logarithms, therefore, it is only necessary to show the mantissas of the logarithms for the figures of the number and the characteristics can be determined by inspection for any positions of the decimal points.

47. Tables of Logarithms. A table of logarithms, or, as it is more frequently called, *logarithm table*, is a table of figures in lines and columns in which the first column contains a list of *numbers* covering a certain range, and the other columns contain the mantissas of the *logarithms* of those numbers in the first column. As the common logarithms of the numbers 1, 10, 100, 1000, etc., have the same mantissas and the characteristics (0, 1, 2, 3, etc.) are determined by inspection, the table can be arranged to give the logarithms of all the numbers from 1 to 100, 10 to 100, 10 to 1000, 100 to 1000, 1000 to 10,000, 1000 to 100,000, etc. Tables arranged in various forms are in use for different purposes.

For very exact work, tables giving the mantissas to 6, 7, 8 or more decimal places are used. For ordinary work in engineering, tables giving the mantissas to four or five decimal places are satisfactory. Such tables are called 4-place tables, 5-place tables, etc., depending on the number of decimal places shown for the mantissas. A 4-place table giving directly the mantissas of the logarithms of numbers 100 to 1000 is given at the back of this book. It is Table IV.

In this table the numbers 10 to 99 are printed in the first column, and in the second column (headed zero, 0) are the mantissas of the logs of these numbers. Since the numbers 10 to 99 have two figures before the decimal point, the characteristic of the logarithm of each is 1. The method of reading the logarithm of any other number than 10–99 is given in the following article.

48. How to Find the Logarithm of a Number in the Table. To find the logarithm of a two-figure number, locate the number in the first column, and on the same line with the number, in the first column at the right, read the mantissa of the logarithm. A decimal point is to be placed before this mantissa. If in the *number* a decimal point follows the first figure, the characteristic is 0 and a 0 is to be written before the mantissa as just read. If a decimal point follows the second figure of the *number*, the characteristic of the logarithm is 1.

To find the logarithm of a three-figure number (regardless of decimal point) locate the first two figures in the first column at the left and the third figure (0 to 9) at the top of the page at the head of one of the other columns. The mantissa is then found on the line with the first two figures and in the column under the third figure. A decimal point is to be placed before every mantissa. The characteristic is a figure or number which is *one less* than the number of figures to the left of the decimal point in the original number whose logarithm was to be found.

As examples, the mantissa of the logarithm of the number 73 is .8633, and the characteristic is 1. Therefore, log 73 = 1.8633. Similarly, log 7.3 = 0.8633; log 730 = 2.8633. As an example of the logarithm of a three-figure number: the mantissa of log 257 is on the line with 25 and in the column headed 7, and is .4099. There are three figures before the decimal point in the number 257 so the characteristic of the logarithm is 2. Therefore, log 257 = 2.4099. But log 25.7 = 1.4099; log 2570 = 3.4099; etc. Similarly, log 47.3 = 1.6749; log 9.28 = 0.9675.

To find the mantissa of the logarithm of a four-figure number, first find the mantissa for the first three figures as just described. To this mantissa add (in the last decimal places) the number found on the same line with the first figures and in the column headed by the fourth figure at the right side of the page. The characteristic is then determined as usual.

Example: To find log 5324. To the first three figures corresponds the mantissa .7259, and on the same line in the column with the fourth figure (4) at the right is found the number 3. The complete mantissa is therefore 7259 + 3 = .7262. Since there are four figures to the left of the decimal point in the number, the characteristic is 3. Therefore, log 5324 = 3.7262. Similarly, for log 672.8 the mantissa is 8274 + 5 = .8279, and the characteristic is 2. Hence, log 672.8 = 2.8279. In the same way, log 12.98 = 1.1134.

The mantissa of the logarithm of a decimal fraction is found as for any other number, ignoring the decimal point in the fraction. The characteristic is then a negative number which is *one more* than the number of ciphers immediately to the right of the decimal point in the decimal fraction. Thus, log .028 = 8.4472 − 10; log .596 = 9.7752 − 10; log .008824 = 7.9457 − 10; etc.

Exercises

Find from Table IV the logarithms of the following numbers:
123; 1234; 72.8; 8.924; .1111; 94560; 3.567; 69.96; .07725; 68; .005783.

49. How to Find the Number Corresponding to a Known Logarithm. Questions like the following arise in the use of logarithms:

The logarithm of a certain number is 2.9325, what is the number? The number is obviously to be found by reversing the procedure of the preceding article. Thus, the mantissa is located in the body of the table, and is found on the line with the figures 85 of the first column and in the column with 6 at the top. The three figures of the required number are therefore 856, and since the characteristic of the logarithm is 2, there are three figures before the decimal point in the required number, which is therefore 856. Therefore, 2.9325 = log 856. The number corresponding to a certain logarithm is called an *anti-logarithm.* The last result would be written anti-log 2.9325 = 856.

From the example just given, therefore, in order to find an anti-logarithm we proceed as follows: If possible locate the *mantissa* of the given logarithm in the body of the table, ignoring the characteristic. The first two figures of the required anti-logarithm are those on the line with the mantissa in the first column, and the third figure is at the head of the column containing the mantissa. When the figures of the number have been found in this way, note the *characteristic* of the given logarithm and point off to the *left* of the decimal point in the anti-logarithm, a number of figures *one more* than the characteristic.

Example: To find anti-log 1.5843. The mantissa 5843 is located on the line with the figures 38 and in the column headed 4. The figures of the required number are therefore 384, and since the characteristic is 1, *two* figures are pointed off at the left, giving 38.4. Therefore, anti-log 1.5843 = 38.4.

When the exact mantissa of a given logarithm cannot be located in the table proceed as follows: Locate the next smaller mantissa and determine the three figures of the anti-logarithm corresponding to this mantissa. Subtract this mantissa from the given mantissa and on the same line locate the difference in one of the columns at the right of the page. The figure at the head of this column is the fourth figure of the required number. The decimal point is located as stated above, by means of the characteristic.

Example: To find anti-log 2.8719. The mantissa 8719 is not in the table but the next smaller in the table is 8716, and this corresponds to the figures 744. The difference in mantissas is 8719 − 8716 = 3, and on the same line, at the right of the page, 3 is found in the column headed 5. The fourth figure is therefore 5 and the required figures are 7445. Since the characteristic of the given logarithm is 2, the number is 744.5, that is, anti-log 2.8719 = 744.5.

If the characteristic of a given logarithm is such a number that more figures are to be pointed off than are obtained from the table, zeros are annexed at the right of the anti-log number to make up the lack. Thus, the figures of anti-log 5.1937 are 1562, but with the characteristic 5 there must be six figures before the decimal point. Two zeros must therefore be annexed, and anti-log 5.1937 = 156200. Similarly, anti-log 7.9463 = 88,360,000.

If the characteristic of a given logarithm is negative, the antilogarithm is a decimal fraction, and there may or may not be zeros immediately at the right of the decimal point in the fraction. The number of these zeros is *one less* than the negative characteristic. The figures are found as usual and the decimal point then located, as determined by the characteristic.

Example: The figures of anti-log 8.1937 − 10 are 1562, and since the characteristic is negative, the number is a decimal fraction. The number of zeros immediately at the right of the decimal point is one less than the negative characteristic 2, that is, one zero; the number is, therefore, .01562. Then, anti-log 8.1937 − 10 = .01562. Similarly, anti-log 9.8719 − 10 = .7445.

Exercises

Find from Table IV the numbers whose logarithms are the following: 1.2504; 2.4031; 0.8368; 3.7806; 4.9720; 9.3885 − 10; 8.2873 − 10; 0.2897; 0.8654.

Chapter 9

USE OF LOGARITHMS IN ARITHMETIC

50. Introduction. One of the chief objects of the study of logarithms is to shorten and simplify the calculations of arithmetic. Logarithms cannot be used in addition and subtraction, but they are very useful in multiplication, division, root extraction and raising numbers to powers. Their use offers little advantage in multiplication and division of small numbers when the quotient is exact, but in involution and evolution and in handling numbers of several figures or inexact quotients they offer the greatest advantages and, in many cases, allow the finding of powers and roots which cannot be found at all by the ordinary methods of arithmetic.

We take up in this chapter the methods of use of logarithms in the various forms of calculation which we have already studied by other methods.

51. Multiplication with Logarithms. From the illustrations and explanations in articles 44 and 45 we have the following rule for

MULTIPLICATION WITH LOGARITHMS: *Find the logarithm of each of the factors and add these logarithms. Find the anti-logarithm of the sum. This anti-logarithm is the desired product.*

As an example, we multiply 28 × 16 = 448 by logarithms.

From the table,

$$\begin{aligned} \log 28 &= 1.4472 \\ \log 16 &= 1.2041 \\ \hline \text{Add} \qquad & \ 2.6513 \\ \text{Anti-log} = \text{Prod.} &= 448 \end{aligned}$$

A convenient form in which to show the work is the following:

$$\begin{aligned} \log 28 &= 1.4472 \\ +\log 16 &= 1.2041 \\ \hline \log \text{Prod.} &= 2.6513 \\ \text{Prod.} &= 448 \end{aligned}$$

Here the plus sign before the second logarithm indicates that it is to be added to the first. When the two logarithms on the right are added to give 2.6513, this being the logarithm of the product is so indicated, and the product itself is the anti-logarithm of this logarithm. The logarithms and anti-logarithm are read from the table in the manner explained in the last chapter.

In multiplying two such small numbers as 16 and 28 there is, of course, no saving of time or trouble in the use of logarithms. In the multiplication of several factors, however, or of factors containing several figures, the use of logarithms frequently requires less time and labor than straight multiplication.

As an example let us multiply 1.623 × 27.31 × 112.4. We have:

$$\begin{aligned} \log 1.623 &= 0.2103 \\ +\log 27.31 &= 1.4364 \\ +\log 112.4 &= 2.0507 \\ \hline \log \text{Prod.} &= 3.6974 \\ \text{Prod.} &= 4982 \end{aligned}$$

Of course, when multiplied out, there will be several decimal places to the right of the decimal point, but as each of the factors is given only to four figures, then, according to the principles and methods of Chapter IV, the result will be known correct only to four figures. The four-place logarithm table has automatically made this adjustment and the last figure (2) in the product is the nearest correct figure.

In multiplying decimals and whole numbers, there will be both positive and negative characteristics to add in the logarithms. The method of performing this operation is illustrated in the following example.

Example: Multiply .0126 × .2831 × 35.6

Let N equal the result of this multiplication

$$N = .0126 \times .2831 \times 35.6$$

$$\log N = \log .0126 + \log .2831 + \log 35.6$$

$$\begin{aligned} \log .0126 &= 8.1004 - 10 \\ \log .2831 &= 9.4520 - 10 \\ \log 35.6 &= 1.5514 \\ \hline \log N &= 19.1038 - 20 \end{aligned}$$

For convenience we write 19.1038 − 20 as 9.1038 − 10

$$\log N = 9.1038 - 10$$
$$N = .1270$$

52. Division with Logarithms. Since division is the operation inverse to multiplication, the rule for division with logarithms is found by reversing that for multiplication. That is, as seen in article 45, for

DIVISION WITH LOGARITHMS: *Subtract the logarithm of the divisor from that of the dividend. The anti-logarithm of the remainder is the quotient.*

As an example, we will divide 928 by 29. We have

$$\begin{array}{lr} & \log 928 = 2.9675 \\ & \log 29 \;\;= 1.4624 \\ \hline \text{Subtracting} & 1.5051 \\ \text{Anti-log} = \text{Quotient} = 32 & \end{array}$$

A simple and convenient form for writing this out is the following:

$$\begin{array}{rl} \log 928 & = 2.9675 \\ -\log 29 & = 1.4624 \\ \hline \log \text{Quot.} & = 1.5051 \\ \text{Quot.} & = 32 \end{array}$$

Here the minus sign before the second logarithm indicates that it is to be subtracted from the first. The Quotient is the anti-logarithm of the number which is indicated as log Quot.

The following example will illustrate how negative characteristics are handled in division.

Example: Divide 81.75 by .0453

Let N equal the result of this division

$$N = 81.75 \div .0453$$
$$\log N = \log 81.75 - \log .0453$$
$$\log 81.75 = 1.9125$$
$$\log .0453 = 8.6561 - 10$$

Since the required subtraction cannot be performed as shown above, we write log 81.75 as 11.9125 − 10 for convenience. We now have

$$\begin{array}{rl} \log 81.75 & = 11.9125 - 10 \\ \log .0453 & = \;\;8.6561 - 10 \\ \hline \log N & = \;\;3.2564 \\ N & = 1805 \end{array}$$

53. Finding Powers with Logarithms. If it is desired to raise a number to a certain power, the usual procedure is to multiply, using the given number as a factor a number of times equal to the exponent. Thus, $12^3 = 12 \times 12 \times 12 = 1728$. To do this by logarithms according to the rule for logarithmic multiplication would require that the logarithm of 12 be set down three times and added. But this gives three times log 12, or $3 \times (\log 12)$. Similarly, to find 127^5 we would get $5 \times (\log 127)$. These illustrations and the discussion of article 45 show that in order

TO RAISE A NUMBER TO ANY POWER: *Find the logarithm of the number, and multiply this logarithm by the exponent. The anti-logarithm of the product is the required power.*

Thus, to find 12^3 by logarithms, we have:

$$\log 12 = 1.0792$$
$$3 \times (\log 12) = 3.2376$$
$$\text{Anti-log } 3.2376 = 12^3 = 1728$$

Similarly, to find 1.396^4,

$$\log 1.396 = 0.1449$$
$$4 \times (\log 1.396) = 0.5796$$
$$\text{Anti-log} = 1.396^4 = 3.799$$

Similarly, to find 139^4,

$$\log 139 = 2.1430$$
$$4 \times (\log 139) = 8.5720$$
$$\text{Anti-log} = 139^4 = 373{,}300{,}000$$

By actual multiplication $139^4 = 373{,}301{,}041$.

Thus, as seen in article 51, a very large number cannot be found exactly with a table of only four places. The error is, in this case, however, only 1041 in a total of over 373 millions, which is relatively very small.

The following example will illustrate how negative characteristics are handled in raising a number to a power.

Example: Find the value of $(.867)^4$

Let N = The required result

$$N = (.867)^4$$
$$\log N = 4 \log .867$$
$$\log .867 = 9.9380 - 10$$
$$4 \log .867 = 39.7520 - 40$$

The logarithm 39.7520 − 40 may be written as 9.7520 − 10 since the characteristic 39. − 40 is equivalent to the characteristic 9. − 10.

Thus $\log N = 9.7520 - 10$
$N = .565$

54. Extracting Roots with Logarithms. Multiplication, division and involution with whole number exponents can be performed in the ordinary regular manner without logarithms, even though the work can often be greatly shortened by their use. On the other hand, many roots cannot be extracted at all by direct calculation, and even in the possible cases, the work is usually very long and laborious, especially for the higher roots which require repeated applications of the square or cube root rules. It is in such cases as these in particular, and indeed in root extraction in general, that the full power and great convenience of logarithms is most clearly seen and highly appreciated.

The rule and the procedure for root extraction with logarithms are simplicity itself, and are immediately obvious. Thus, since evolution is the operation inverse to involution, the procedure is exactly the opposite of that for powers, as already illustrated in article 45. Therefore,

TO FIND ANY ROOT OF A NUMBER: *Find the logarithm of the number, and divide this logarithm by the root index. The anti-logarithm of the quotient is the desired root.*

For example, to find the square root of 1849: We have

$$\begin{aligned} \log 1849 &= 3.2669 \\ (\log 1849) \div 2 &= 1.6335 \\ \text{Anti-log} = \text{Root} &= 43 \end{aligned}$$

As a second example, to find $\sqrt{59536}$:

$$\begin{aligned} \log 59536 &= 4.7748 \\ (\log 59536) \div 2 &= 2.3874 \\ \text{Anti-log} = \text{Root} &= 244 \end{aligned}$$

The two examples given here are those which are worked out in article 39 by the ordinary rule. Comparison with the solutions there given shows the great advantage of the logarithmic method.

As another example, let us find $\sqrt[3]{78347809.639}$. In a four-place table, of course, log 78347809.639 cannot be found exactly but it is very nearly the same as log 78350000. We have, therefore,

$$\log 78347809.639 = 7.8941$$
$$(\log 78347809.639) \div 3 = 2.6314$$
$$\text{Anti-log} = \text{Root} = 427.9$$

This is the example worked out in article 40. Comparison shows the very great superiority of the logarithmic method.

Let us next find $\sqrt[5]{4594014}$. We have

$$\log 4594014 = 6.6622$$
$$(\log 4594014) \div 5 = 1.3324$$
$$\text{Anti-log} = \text{Root} = 21.5$$

This root cannot be found at all by the ordinary methods. As a check on the exactness of the result; if 21.5 be raised to the fifth power by repeated multiplication the result is found to be

$$21.5^5 = 4594013.84375$$

We will now find the sixth root of 2985984.

$$\log 2985984 = 6.4751$$
$$(\log 2985984) \div 6 = 1.0792$$
$$\text{Anti-log} = \text{Root} = 12$$

To find the sixth root of any number by the ordinary methods requires the successive application of the rules for square root and cube root. Comparison of the logarithmic calculation just performed with the rules and procedure of articles 39 and 40 shows the immense superiority of the logarithmic method. In fact, there can hardly be any comparison.

As a final example we will work out an example involving several operations.

Example: Find the value of $\dfrac{687 \times \sqrt[3]{58.3}}{(7.95)^2}$

Let N = The required result

$$N = \frac{687 \times \sqrt[3]{58.3}}{(7.95)^2}$$

$$\log N = \log 687 + \tfrac{1}{3} \log 58.3 - 2 \log 7.95$$

$$\log 687 = 2.8370$$
$$\log 58.3 = 1.7657$$
$$\log 7.95 = 0.9004$$

$$\begin{aligned} \log 687 &= 2.8370 \\ \tfrac{1}{3}\log 58.3 &= 0.5886 \\ \hline &\ \ 3.4256 \\ 2\log 7.95 &= 1.8008 \\ \hline \log N &= 1.6248 \\ N &= 42.2 \end{aligned}$$

In addition to such roots as those just mentioned it is also necessary, in various parts of applied mathematics, to extract roots of fractional index, as the $\frac{7}{10} = .7$ root, or the 2.63 root; and also to raise numbers to powers of fractional exponent, as the $\frac{4}{5} = .8$ power, the $\frac{5}{2} =$ power (in the study of fluid mechanics), the 1.6 power (in the study of electricity and magnetism). It is also found necessary to calculate roots and powers with *negative* indexes and exponents. In order to perform such calculations as these it is necessary to know more of the underlying principles of logarithms and of general properties of numbers. These are studied in algebra, a subject which should follow arithmetic.

55. The Slide Rule. The *slide rule* is an instrument by means of which solutions may be obtained mechanically for problems which involve multiplication, division, powers, roots, proportion, measurement, etc. It consists of an ordinary wooden rule or ruler about eleven inches long, one and a quarter inches wide, and three eighths inch thick. Besides an ordinary 10-inch rule along the beveled edge divided into quarters, eighths and sixteenths of an inch, it has on the flat face several additional scales which are specially ruled and marked to represent the logarithms of numbers.

In this face of the rule is cut a slot or groove which extends throughout the length of the rule, about five eighths inch wide and three sixteenths deep. In this groove slides a snugly fitting wooden strip which forms a second rule of the same length, with its upper face flush with that of the main rule. This sliding rule is divided and marked like the main rule, and by setting it so that certain numbers on its scales are lined up with certain other numbers on the scales of the main rule, the *logarithms* of those numbers may be added, subtracted, multiplied or divided so that the numbers themselves are, according to the principles of logarithms, multiplied, divided, raised to powers or have roots extracted.

When the markings and reading of the scales are understood, the operation of the slide rule is very simple and all ordinary calculations are performed very rapidly, without writing, and with the accuracy of a three- or four-place table of logarithms.

There are several different forms of the slide rule in general use, all of which are based on the same principles, but the detailed description and explanation of any of them are beyond the scope and plan of this book.*

56. Historical Note. We have seen how many of the ordinary arithmetical calculations may be greatly shortened and simplified by the use of logarithms, especially when the numbers involved, be they whole numbers or fractions, consist of many figures. In surveying, navigation, astronomy, and many branches of science, engineering and business such numbers and calculations occur very frequently. In the early history of mathematics all such calculations were carried out by the ordinary methods and required enormous amounts of time and labor.

It was not until about the year 1614 that the logarithmic method of calculation was invented. Logarithms were invented by John Napier, Baron of Merchiston in Scotland. The base he used was not the number 10 and his tables were not so simple and easy to use as are the tables of the present time. About 1617 the number 10 was adopted as base by William Briggs, Professor of Mathematics at Oxford University in England, who was a friend of Napier. Briggs calculated the first table of common logarithms and since his day all tables of common logarithms have been based on his tables. In honor of him common logarithms are often called *Briggs* logarithms.

About a hundred years after Napier's invention, it was discovered that many of the formulas and much of the theory of the higher mathematics are greatly simplified by the use of logarithms whose base is the decimal number 2.71828+ and logarithms to this base are called *natural logarithms* in distinction from the decimal common logarithms. This number is closely related to the base which Napier used originally and in his honor the natural logarithms are called also *Napierian* logarithms.

* *Note.*—Readers interested in the slide rule are referred to the author's "Manual of the Slide Rule," published by D. Van Nostrand Company, New York. It gives complete descriptions, explanations and instructions for the use of all the standard forms of the slide rule.

About the year 1620 Edmund Gunter in England showed that logarithmic calculations can be carried out mechanically by laying off on a ruler lengths representing the logarithms of the numbers marked on the scales, and then combining these by means of the compasses or dividers. About 1630 Rev. William Oughtred of London perfected these scales and arranged two of them to lie along one another and to *slide* one upon the other so as to make the dividers unnecessary and the calculations much more simple and rapid. He thus invented the slide rule, which is now, therefore, more than three hundred years old.

After Oughtred's time, slide rules of many different forms were designed for different purposes. In the year 1851, Lieut. Amadée Mannheim, a French artillery officer stationed at the great fortress of Metz in Alsace-Lorraine finally gave to the slide rule the form of scales and construction which is now standard in almost the entire world. The standard rule is now generally referred to as the Mannheim Slide Rule. Several modifications of the Mannheim rule have been developed in the United States in recent years and are very widely used.

A fairly complete history of logarithms and the slide rule is given in the book referred to at the end of article 55.

57. Exercises.

In the following exercises use Table IV and follow the rules and methods given in this chapter.

1. Multiply by means of logarithms: 28×36; 28.3×36.3; 765×1.125; $14.68 \times .2685$; $.963 \times .1724$; 1234×4321.

2. Divide by means of logarithms: $685 \div 25.6$; $.7842 \div .3631$; $84760 \div 376.5$; $14.76 \div 28.3$; $1875 \div 3263$; $32630 \div 1.875$.

3. By means of logarithms raise the following numbers to the indicated powers: 48^2; 6.27^3; 14^4; 928^2; $183^{2.4}$; $946.7^{1.6}$.

4. By means of logarithms extract the indicated roots of the following numbers: $\sqrt{9216}$; $\sqrt[3]{9596000}$; $\sqrt[5]{7776}$; $\sqrt[7]{2187}$; $\sqrt[2.5]{7284}$; $\sqrt[1.6]{96.34}$.

5. Perform the following indicated calculations by means of logarithms: $6.85 \times 172.8 \div 43.9$; $.892 \times 46.5 \times 3.76$; $632.5 \times 28.56 \div 154 \times 307.5$; $\dfrac{890 \times 12.34 \times .0637}{87.35 \times 2.274}$.

Chapter 10

RATIO AND PROPORTION

58. Meaning of Ratio. An expression like $\frac{3}{4}$, $\frac{6}{2}$, $\frac{3}{2}$, is usually called a fraction and we have studied the properties of fractions in a previous chapter. We have seen, however, that the *value* of such an expression can be expressed as a decimal fraction, a whole number or a decimal number. Thus, $\frac{3}{4} = .75$, $\frac{6}{2} = 3$, $\frac{3}{2} = 1.5$.

The single number which represents the value of such an expression is called the *ratio* of the two numbers which make up the fraction. Thus $\frac{6}{2} = 3$ is called the "ratio of 6 to 2."

Instead of the fraction line (—) or the division sign (÷) the sign (:) is sometimes used to indicate a ratio. It may be thought of as the division sign (in which the dots represent the positions of the numbers) with the fraction line removed. The fraction $\frac{6}{2}$, or the quotient $6 \div 2$, is then written as the ratio 6 : 2.

Other examples of ratios are: the ratio of 4 to 2, 4 : 2; the ratio of 1 to 3, 1 : 3; the ratio of 21 to 7, 21 : 7; etc. The values of these ratios are, respectively, 2, .333+, 3.

From the examples and definition given, it may be said, in a general way, that the ratio of one number to another represents or indicates the relative value or magnitude of the two numbers, the value of one of them as compared with the other. Thus, in the ratio of 6 to 2, written 6 : 2, the 6 is three times as great as the 2.

59. Meaning of Proportion. When two ratios are equal, the four numbers forming the ratios are said to form a *proportion*, or to be *in proportion*. Thus, the ratio of 21 to 7 is 3, $\frac{21}{7} = 3$; also ratio of 6 to 2 is 3, $\frac{6}{2} = 3$. Therefore, the two ratios are equal, $\frac{21}{7} = \frac{6}{2}$, and the numbers 21, 7, 6, 2 are in proportion.

Using the method of writing used in article 58, the proportion $\frac{21}{7} = \frac{6}{2}$ is sometimes written 21 : 7 = 6 : 2, which is read "21 is to 7 as 6 is to 2."

If any three numbers of a proportion are known, but the fourth unknown, the fourth may be represented by a letter, as in article 4. This may be the initial letter of the word which names the thing the number is to represent, as L for length, W for weight, P for price, etc., or for any number in general the letter x may be used. Thus, suppose one of the two ratios which form a proportion is 22 : 11 or $\frac{22}{11}$, and the first number of the other ratio is 10 but the remaining number is not known. If this number is represented by x the ratio can be written $10 : x$ or $\dfrac{10}{x}$, and the complete proportion is then $\dfrac{22}{11} = \dfrac{10}{x}$, or $22 : 11 :: 10 : x$, "22 is to 11 as 10 is to x." By rules which shall be given presently, the value of x in this proportion can be found so that the ratio of 10 to x will be the same as that of 22 to 11.

From the definition and illustrations given in this article we may say, in general, that, when four numbers are in proportion, the relation between two of them is the same as the relation between the other two, that is, the first is as many times as great (or small) as the second, as the third is times the fourth.

60. The Fundamental Rule of Proportion. In a proportion such as $21 : 7 :: 6 : 2$ the first and last or "end" numbers are called the *extremes* of the proportion and the second and third or "intermediate" numbers are called the *means.* In the proportion given, the extremes are 21 and 2 and the means are 7 and 6. These names will be used regularly in what follows.

It is to be noticed that the product of the means in the proportion $21 : 7 :: 6 : 2$, which is $7 \times 6 = 42$, is the same as that of the extremes, $21 \times 2 = 42$. Similarly, since $4 \div 2 = 2$ and also $6 \div 3 = 2$, we can write $4 : 2 = 6 : 3$, and here also the product of the means, $2 \times 6 = 12$, is the same as that of the extremes, $4 \times 3 = 12$. This is true of any proportion, as can be seen at once by writing any two equal fractions in the form of a proportion and forming the products of the means and of the extremes. We can state, therefore, that in any proportion

The product of the means equals the product of the extremes.

This is the *fundamental rule* of proportion.

61. Solution of a Proportion. The four numbers which form a proportion, the two means and two extremes, are together called the members or *terms* of the proportion. The extremes are the first and fourth terms, and the means the second and third terms.

If any term of a proportion is not known, but the other three are known, then, as has already been stated, the fourth term can be found. When the value of an unknown term of a proportion has been found by any means the proportion is said to have been *solved*, and the process of finding the value of the unknown term is called the *solution* of the proportion.

The solution of a proportion depends on the fundamental rule of proportion given in article 60. In order to solve any proportion, *three* of its terms must be known, and also when *any three* numbers are related to one another as terms of a proportion a fourth number can always be found which is related to those *three* as the fourth term. Problems involving such related numbers are common in arithmetic.

Consider the proportion given in article 59, with the three known and one unknown terms, $22 : 11 :: 10 : x$, the letter x representing the unknown term. Now, according to the fundamental rule, the product of the extremes, $22 \times x$, must be equal to the product of the means, $11 \times 10 = 110$. That is, $22x = 110$, 22 times x equals 110. The number x must, therefore, be 5 and the proportion is $22 : 11 :: 10 : 5$, the solution of the proportion being $x = 5$.

Here the known means were multiplied, and the product was divided by the known extreme, the quotient being the desired unknown extreme, x.

Suppose it is one of the means which is unknown in a proportion. Thus, if $42 : 7 :: x : 5$, what is x? Here again, according to the fundamental rule, $7 \times x = 42 \times 5$. That is, $7x = 210$, and therefore $x = 210 \div 7 = 30$. In this case, the product of the known extremes is divided by the known mean to find the unknown mean, the fourth term of the proportion.

The two examples just given and a study of the fundamental rule show that any proportion, of which three terms are known, can be solved for the fourth by the following

RULE: (i) *If either extreme of a proportion is unknown, divide the product of the means by the known extreme; the quotient is the desired unknown term.*

(ii) *If either mean is unknown, divide the product of the extremes by the known mean; the quotient is the desired unknown term.*

62. Mean Proportional. If, in any proportion, the means are equal, that is, both means are the same number, then that number is called a *mean proportional* to the other two, or the mean proportional of the other two. It is also sometimes called the *geometric mean* for the reason which we shall see in article 108.

As an example, in the proportion $8 : 4 :: 4 : 2$, 4 is the mean proportional of 8 and 2. Similarly, in the proportion 81 : 27::27 : 9, 27 is a mean proportional to 9 and 81. In such a case, the product of the means is the mean proportional multiplied by itself, that is, the square of the mean proportional. The square of the mean proportional is, therefore, by the fundamental rule, equal to the product of the other two numbers, the extremes. The mean proportional is, therefore, the *square root* of the product of those two. This result is concisely stated as the

RULE: *The mean proportional of any two numbers is the square root of their product.*

The examples already given will illustrate this rule. Thus, in the proportion 8 : 4::4 : 2, $4 = \sqrt{(8 \times 2)} = \sqrt{16}$; in the proportion 81 : 27::27 : 9, $27 = \sqrt{(9 \times 81)} = \sqrt{729}$.

If any two numbers are given and their mean proportional is required, as for example 3 and 12, we can write $3 : x::x : 12$, x being the unknown mean proportional. Then, $x = \sqrt{(3 \times 12)} = \sqrt{36} = 6$. What is the mean proportional of 2 and 128? By the rule, it is $x = \sqrt{(2 \times 128)} = \sqrt{256} = 16$.

63. Direct and Inverse Proportion. Let us consider the following problem: If 6 tons of coal cost 144 dollars, how many tons can be bought for 72 dollars? A little consideration will show that the answer to the equation is 3 tons. But how can the problem be stated in terms of arithmetical symbols so that a direct calculation will give the required result? The answer to this question will reveal the utility and the real meaning of the principles of proportion, and in addition a second form of statement of a proportion called *inverse* proportion.

The general solution of the problem is something like this: 6 tons cost 144 dollars, so one ton costs $144 \div 6 = 24$ dollars; therefore, the number of tons which can be bought for 72 dollars is $72 \div 24 = 3$ tons. This means that \$144 ÷ 6 tons = 24 dollars per ton must be the *same* as \$72 ÷ (the unknown number of tons) = 24 dollars per ton, or if x represent this unknown number, then $144 \div 6$ is the *same* as $72 \div x$. That is, the *ratio* of 144 to 6 or $\frac{144}{6}$, equals the ratio of 72 to x or $\frac{72}{x}$.

This statement is written $\frac{144}{6} = \frac{72}{x}$. This is a proportion; in the usual arithmetic form it is written $144 : 6 = 72 : x$. According to the rule of article 61 the value of the unknown x is then $x = (6 \times 72) \div 144 = 3$ as already seen. The original problem is thus stated as a proportion and is solved by solving a proportion.

In writing a proportion it is better to put numbers representing the same things in the same ratio. Thus, instead of saying that (\$144) : (6 tons) = (\$72) : (x tons), "dollars are to tons as dollars are to tons" it is better to say "dollars are to dollars as tons are to tons": We would then write (\$144) : (\$72) = (6 tons) : (x tons), or simply $144 : 72 = 6 : x$, it being understood that the ratio 144 : 72 refers to dollars and the ratio $6 : x$ to tons. The solution is the same as before: for, $144 \times x$ must equal 72×6, and hence, $x = (72 \times 6) \div 144 = 3$.

The operation just carried through is an example of the method of solving problems by means of proportion; this general question will be discussed further in article 64. There is another question, however, which occurs in connection with this problem. This matter will next be discussed.

In this problem \$144 bought 6 tons and \$72 bought 3 tons; also \$240 would buy 10 tons. In other words, more money would buy more coal; less money, less coal. This may be expressed by saying that "more requires more and less requires less." In such a case it is said that the amount of coal purchased is in *direct proportion* to the amount of money spent, the amount of the one thing involved in the problem *increases* as the other also increases. All the preceding discussions of this chapter apply to direct proportion.

Suppose now that in a certain problem two quantities are so related that the one *decreases* as the other increases; in other words, "more requires less and less requires more." In this case the amounts or quantities of the two things are said to be in *inverse proportion.*

An example of inverse proportion is the following: If 3 men can do a certain job in 6 days, how many days will be required for 5 men to do the same job, all working at the same rate as before? To work this out in step-by-step detail, the process is somewhat long and involved. The problem can be solved by means of inverse proportion, however, very simply and directly.

Since it will require less time for the 5 men than for the 3 men, we cannot write *directly* that (3 men) : (5 men) = (6 days) : (x days), but must remember that "more requires less" and invert the order of writing the number of days. We, therefore, write inversely, that (3 men) : (5 men) = (x days) : (6 days), or simply $3 : 5 = x : 6$, remembering to write "men to men as days to days" but writing the numbers in one ratio in the inverse order of the corresponding numbers in the other ratio.

Once having written the inverse proportion $3 : 5 = x : 6$, the solution is carried out by the regular rule for the proportion *as written;* the solution is $x = (3 \times 6) \div 5 = 3\frac{3}{5}$ days.

An easy way to remember how to write direct and inverse proportion is to write out the ratios in the form of fractions, inverting one of the fractions in inverse proportion. Thus, in the first example of this article, "dollars to dollars as tons to tons," the ratios would be written $\frac{144}{24}$ and $\frac{6}{x}$ and the proportion $\frac{144}{24} = \frac{6}{x}$. This is then put in the usual form $144 : 72 :: 6 : x$ and solved as usual. Similarly, in the second example, "men to men as days to days," the ratios are $\frac{3}{5}$ and $\frac{6}{x}$; inverting one of these, say the second, we have $\frac{3}{5}$ and $\frac{x}{6}$. The proportion is then $\frac{3}{5} = \frac{x}{6}$, or $3 : 5 :: x : 6$, and the solution is, as usual, $x = (3 \times 6) \div 5 = 3\frac{3}{5}$.

64. Solution of Problems by Proportion. The discussion of direct and inverse proportion in the preceding article has brought out the meaning of proportion and indicated the two methods of solving problems by the use of proportion. A very large part of the problems of ordinary arithmetic can be solved by means of proportion, and in all these the rules of article 61 are used when the problem is once *formulated* (formed and written out). In the formulation, however, care must be taken to determine which are the related quantities or numbers, whether the relation is direct or inverse, and which is the unknown required quantity or number. In this article we give a few suggestions as to procedure and illustrate the methods by solving a few simple problems.

Referring to the examples of article 63 it is seen that in each case the procedure was somewhat as follows: We

(i) Determine the two quantities of the same kind;

(ii) Determine the quantity which is of a different kind and let x represent a second quantity of this same kind;

(iii) Form the ratio of the first two quantities, and write this ratio as the first of a proportion;

(iv) Determine from the conditions of the problem whether "more requires more and less requires less," or the reverse;

(v) If "more requires more," form the ratio of the two quantities of step (ii) in the same order as in (iii); if "more requires less, or less requires more," use the inverse order;

(vi) Write the final ratio of (v) as the second ratio of the proportion of (iii);

(vii) Solve the proportion thus formed, using the rules of article 61.

This procedure is sufficient for the solution of any problem which can be solved by direct or inverse proportion. We give below a few illustrative solutions.

Problem 1. If 9 bushels of wheat make 2 barrels of flour, how many barrels will 100 bushels make?

Solution. Here we are to write bushels to bushels as barrels to barrels, and as "more requires more," the proportion is direct. The first ratio, involving bushels, is $9 : 100$. Letting x represent the required number of barrels, the direct ratio involving barrels is $2 : x$. The proportion is therefore $9 : 100 = 2 : x$, and by the rule the solution is $x = (2 \times 100) \div 9 = 22\frac{2}{9}$ barrels.

Problem 2. If 20 men can do a job in 15 days, how many more men will be required to do it in four fifths of that time?

Solution. The number of days which is $\frac{4}{5}$ of 15 days is 12. We must, therefore, find how many men are required to do the job in 12 days and then see how many more than 20 this number is.

The first ratio, involving days, is $15 : 12$, and since less time will require more men the second ratio, involving men, is in the inverse order, $x : 20$. The proportion is, therefore, $15 : 12 = x : 20$, and hence $x = (20 \times 15) \div 12 = 25$ men, which is 5 *more* than 20.

Problem 3. Two numbers are to each other as 5 to $7\frac{1}{2}$ and the smaller is 164.5; what is the greater?

Solution. The first ratio is $5 : 7\frac{1}{2}$, or decimally, $5 : 7.5$; and since the unknown number x is the greater, the ratio of the two numbers is $164.5 : x$. These ratios being the same the proportion is $5 : 7.5 = 164.5 : x$. Therefore, $x = (7.5 \times 164.5) \div 5 = 246.75$.

Problem 4. The force applied to one end of a lever to lift a certain weight at the other end is related to that weight inversely as the distances of force and weight from the pivot or support point of the lever. If a lever 8 feet long is supported 4 inches from one end, what weight can be lifted at the end nearest the support by a force of 50 pounds applied at the other end?

Solution. The statement that the force and weight are inversely proportional to their distances means that the lesser force is at the greater distance from the support. The distance of the support from one end being 4 inches, the distance of the other end is 7 feet 8 inches, or 92 inches, the full length being 8 feet. The ratio of the distances is, then, 92 : 4. The ratio of the corresponding force and weight in pounds is in the inverse order, $x : 50$, x being the greater unknown weight at the 4-inch distance. The proportion is, therefore, $92 : 4 = x : 50$. Hence $x = (50 \times 92) \div 4 = 1150$ pounds.

65. Exercises and Problems.

Solve the following proportions for the unknown term, denoted by x:

1. $6 : 13 = 9 : x$. 2. $3 : 5 = x : 15$. 3. $6 : x = 17\frac{1}{2} : 52.5$.
4. $x : 175 = 7 : 5$.

5. If 7 men's shirts cost $29.75, how many men's shirts can be bought for $68?

6. If a speculator gains $26.32 by investing $325, how much would he gain by investing $2275?

7. A dealer can buy 18 dozen eggs for $9.72. How many dozen eggs can he buy for $12.42?

8. If a rod 4 feet long casts a shadow 7 feet long, what is the height of a building which at the same time casts a shadow of 198 feet?

9. A tree which is known to be 75 feet high stands on one bank of a river at the water's edge and casts a shadow which just reaches straight across to the other bank, and at the same time a yard stick casts a 4-foot shadow. Find the width of the river.

10. A lever is 6 feet long and is supported 6 inches ($\frac{1}{2}$ foot) from one end. How much weight at that end can be raised by a 50-pound force at the other end?

11. The hook supporting a load on a scale is $\frac{1}{2}$ inch from the pivot on which the scale beam is supported, and the sliding weight is 5 pounds. How far from the pivot must the sliding weight be to balance a load of 200 pounds?

12. At a certain time of day when a vertical yard stick casts a shadow 19 inches long, one of the giant redwood trees of California casts a shadow 213 long. How high is the tree?

13. A soldier in Egypt noted that at the same time when his cane placed upright cast a shadow 26 inches long, the Great Pyramid of Cheops cast a shadow about 153 paces long, stepped off from the center of the base. If the cane was 34 inches long and his pace about $2\frac{1}{2}$ feet, how high is the pyramid? (It is said that the ancient Greek Philosopher and business man Thales first

used this method to measure the heights of the pyramids when he went to Egypt on a vacation.)

14. A motorist finds that on a certain test trip $4\frac{1}{2}$ gallons of gasoline carried him 63 miles. Without calculating the mileage per gallon, find directly by proportion how much gasoline will be required to take him on a 675-mile trip, traveling under the same conditions.

15. Four soldiers dig a trench of a certain depth and width and 30 feet long in half an hour. How many soldiers digging at the same rate will be required to extend the same trench (in the same kind of soil) to a total length of 55 feet in 20 minutes more?

Chapter 11

SERIES AND PROGRESSIONS

66. Meaning of a Series. As understood in arithmetic, a *series* is a list or succession of numbers arranged in a certain order and so related that each number is formed in the same manner from the preceding number in the list. In counting, the ordinary numbers 1, 2, 3, 4, 5, etc., form a series in which each number is formed by adding 1 to the preceding number. Similarly, in counting by twos, 2, 4, 6, 8, 10, 12, etc., we have a series in which the difference between consecutive numbers is 2. The list of numbers in article 44 is formed by *multiplying* the preceding number by 2.

The numbers which form the series are called the *terms* of the series. If a series has a definite number of terms, that is, it does not run on indefinitely both ways, the first and last numbers are called the *first term* and the *last term.*

The method or rule by which each term is formed from the preceding term is called the *law of the series.*

The sum of all the numbers or terms of a series is called the *sum of the series.*

Suppose that in a certain lottery or raffle, the tickets are numbered consecutively 1 to 100 and the cost of each in cents is the number printed on it. What are the proceeds of the raffle? Again, what is the sum of the first ten numbers of the series of article 44?

Suppose that it is desired to raise a specified sum of money by means of a lottery such as that just described. How can we calculate the necessary number of tickets? And again, without referring to the list of numbers in article 44, how can we tell the number of the term which is equal to 4096?

By means of the principles and rules of series, the questions just asked, as well as other similar or related questions, can readily be answered. As seen from the examples of series given above, however, there are two kinds of series. We proceed to consider these separately and in some detail.

67. Arithmetical Progression. If each term in a series is formed by *adding* a certain number, and always the same number, to the preceding term, the terms of the series are said to be in *arithmetical progression.* The terms of the series are called the *terms* of the progression, and so also for the *sum.* The number added to each term to form the next, that is, the difference between consecutive terms, is called the *common difference* of the series. The law of arithmetical progression is, therefore, *addition of the common difference.*

The important parts or *elements* of an arithmetical progression are therefore the

1. *First* term;
2. *Last* term;
3. *Number* of terms;
4. Common *difference;*
5. *Sum* of the series.

These five elements are related and connected by important rules such that if any three are known the remaining two can be calculated.

68. Rules of Arithmetical Progression. Consider the arithmetical progression 5, 7, 9, 11, 13, etc., and suppose that there are 27 terms: What is the last term? Here the first term is 5 and the common difference is 2. Since there are 26 terms after the first, that is, one less than the total number of terms, the common difference is added 26 times to form the last term, which is therefore $5 + (26 \times 2) = 5 + 52 = 57$.

In this case, there were given the first term, the common difference and the number of terms, and it was required to find the last term. It is seen at once that the method used will apply to any similar case, and from the solution we, therefore, have the following:

RULE I. *The last term of an arithmetical progression is equal to the first term plus the product of the common difference by the number of terms less* 1.

Reversing this rule we have immediately

RULE II. *The first term of an arithmetical progression is equal to the last term minus the product of the common difference by the number of terms less* 1.

If we let a represent the first term, l the last term, n the number of terms and d the common difference, these two rules are very neatly and concisely expressed by writing

$$\text{I.}\quad l = a + [(n - 1) \times d].$$

$$\text{II.}\quad a = l - [(n - 1) \times d].$$

By means of these rules, either extreme can be found when the other extreme, the common difference, and the number of terms are known.

From either rule it is seen that the difference between the extremes is the product of the common difference by the number of terms less 1. If, therefore, the extremes and the number of terms are known, the common difference is found by

RULE III. *The common difference of an arithmetical progression is found by dividing the difference of the extremes by the number of terms less* 1.

Similarly, the number of terms less 1 is the difference of the extremes divided by the common difference of the progression. From this relation, therefore, we have

RULE IV. *To find the number of terms in an arithmetical progression, divide the difference of the extremes by the common difference of the progression and to the quotient add* 1.

Using the abbreviations already used in connection with Rules I and II, Rules III and IV may very conveniently be abbreviated as follows:

$$\text{III. } d = (l - a) \div (n - 1).$$

$$\text{IV. } n = [(l - a) \div d] + 1.$$

The four rules so far given enable us to find any one of the first four elements of an arithmetical progression tabulated in article 67 when the other three of the four are known. Let us see now how the fifth element, the *sum*, is related to the other four.

Consider an arithmetical progression, such as 2, 5, 8, 11, 14, which has 5 terms, extremes 2 and 14, and common difference 3. Writing the sum of these terms in both direct and inverse order, we have

$$\begin{array}{lrl} & 2 + 5 + 8 + 11 + 14 &= 40, \text{ the sum.} \\ & 14 + 11 + 8 + 5 + 2 &= 40, \text{ the sum.} \\ \hline \text{Add,} & 16 + 16 + 16 + 16 + 16 &= 80, \text{ twice the sum.} \end{array}$$

From this it is seen that 16, which is the sum of the extremes, $2 + 14$, multiplied by 5 equals 80, which is twice the sum of the series. Therefore, the sum is $40 = 80 \div 2 = (5 \times 16) \div 2 = 5 \times (2 + 14) \div 2$. Since the same method will give the sum of any such progression, we can state the method as

RULE V. *The sum of the terms of an arithmetical progression is the product of half the number of terms by the sum of the extremes.*

If we let S represent the sum and use the same other abbreviations as before the rule is very simply expressed as follows:

$$\text{V. } S = \tfrac{1}{2}n \times (a + l).$$

By using the various rules in succession or in combination any problem involving an arithmetical progression may be solved. For example, from Rule V the number of terms can be found from the sum and extremes by doubling the sum of the progression and dividing by the sum of the extremes. Knowing then the extremes and the number of terms the common difference can be found by Rule III. Similarly, other combinations can be worked out by the reader.

As an example, suppose that it is desired to raise $9.30 by a raffle like that described in article 66, the lowest ticket being numbered 2 and the highest 60. How many tickets will be necessary and what will be the common difference between the consecutive ticket numbers?

Here we have $a = 2$, $l = 60$, $s = 930¢$. By Rule V, therefore, the number of terms (tickets) is $n = (2 \times 930) \div (2 + 60) = 1860 \div 62 = 30$. By Rule III, the common difference is then $d = (60 - 2) \div (30 - 1) = 58 \div 29 = 2$. There will therefore be 30 tickets numbered 2, 4, 6, 8 and so on to 60, the sum of all the numbers being 930.

69. Geometrical Progression. If each term of a series is formed by *multiplying* the preceding term by a certain number, and always the same number, the numbers or terms of the series are said to be in *geometrical progression*, and the series itself is called a geometrical series or a geometrical progression. The *extremes*, or first and last terms, the *number* of terms, and the *sum* of the progression or series, all have the same meanings as in an arithmetical progression. The number by which each term is multiplied to form the next term is called the common multiplier, or more often the *common ratio*. This name is used because, from the law of the series, the quotient of each term by the preceding, or *ratio* of each term to the preceding, is the same for all. The law of the series is therefore *multiplication by the common ratio*.

If the common ratio is a number greater than 1, each term is obviously greater than the preceding term, and the series or progression is called an *increasing* series. If the ratio is a number less than 1, that is, a fraction, each term is obviously less than the preceding, and the series is called a *decreasing* series.

If the number of terms in a decreasing series is infinitely great, the last term is obviously infinitely small, that is, zero. In this case the series is called an *infinite series.* The sum of an infinite series of this kind is not infinitely great but has a very definite value and in the higher mathematics it is a very important quantity.

The series of numbers in article 44 is a geometrical progression in which the first term is 1 and the common ratio is 2. The value of the last term depends on the number of terms used and the series can be extended indefinitely. As there given, the number of terms is 21 and the last term is 1,048,576.

From the definitions, descriptions and example here given we have as the five *elements* of a geometrical progression the

1. *First* term;
2. *Last* term;
3. *Number* of terms;
4. Common *ratio;*
5. *Sum* of the series.

These five elements are related and connected by a set of rules such that when any three are given the remaining two can be calculated. These rules are given in the next article.

70. Rules of Geometric Progression. If the first term, the common ratio and the number of terms of a geometrical progression are known the last term is easily found. Thus, take a geometrical progression with first term 3 and ratio 2, and consider 5 terms. These terms are,

$$3,\ 3 \times 2,\ (3 \times 2) \times 2,\ (3 \times 2 \times 2) \times 2,\ (3 \times 2 \times 2 \times 2) \times 2,$$

or, counting the number of times 2 is used as a factor in each,

$$3,\ 3 \times 2,\ 3 \times 2^2,\ 3 \times 2^3,\ 3 \times 2^4.$$

Similarly, if there are 8 terms they are,

$$3,\ 3 \times 2,\ 3 \times 2^2,\ 3 \times 2^3,\ 3 \times 2^4,\ 3 \times 2^5,\ 3 \times 2^6,\ 3 \times 2^7,$$

and in the same way the series could be written out for any number of terms or any common ratio.

Now it is to be noted in each of these that the last term is equal to the first multiplied by a certain power of the common ratio, namely the power whose exponent is one less than the number of terms. This is true in any case and gives us

RULE I. *The last term of a geometrical progression equals the first term multiplied by the ratio raised to a power one less than the number of terms.*

Reversing this rule we have at once

RULE II. *The first term of a geometrical progression equals the last term divided by the ratio raised to a power one less than the number of terms.*

These two rules enable us to find either of the extremes when we know the other extreme, the ratio and the number of terms.

If we let a represent the first term, l the last term, n the number of terms, and r the common ratio, these two rules can be very neatly and concisely expressed as follows:

$$\text{I. } l = a \times r^{n-1}.$$

$$\text{II. } a = l \div r^{n-1}.$$

If both extremes and the number of terms are known, then, since the last term is equal to the first term multiplied by a certain power of the ratio, if the last term be divided by the first, the quotient will be that power of the ratio, and in order to find the ratio it is only necessary to extract the corresponding root of that quotient. This result is expressed in

RULE III. *Divide the last term by the first, and of the quotient extract the root whose index is the number of terms less* 1. *The result is the common ratio.*

Using the same abbreviations as before, this rule can be expressed in symbols as follows:

$$\text{III. } r = \sqrt[(n-1)]{(l \div a)}.$$

The powers and roots involved in these three rules are most conveniently calculated by means of logarithms as in articles 53 and 54.

When the extremes and ratio are known, the number of terms may be found by either of two rules, one involving repeated division, and the other involving logarithms. These are:

RULE IV*a*. *Divide the last term by the first, divide this quotient by the ratio, the quotient thus obtained by the ratio again, and so on in successive division by the ratio, until the quotient is* 1. *The number of times the ratio is used as a divisor, plus* 1, *is the number of terms.*

RULE IV*b*. *From the logarithm of the last term, subtract the logarithm of the first term, and divide the remainder by the logarithm of the ratio. The quotient plus* 1 *is the number of terms.*

Rule IV*a* cannot conveniently be written in an abbreviated form. Using the same symbols as before, however, Rule IV*b* can be expressed as follows:

$$\text{IV. } n = [(\log l - \log a) \div (\log r)] + 1.$$

The sum of a geometrical progression can be found when any three of the other elements are known. We shall derive two rules for finding the sum in different cases.

Consider the 4-term progression with first term 5 and ratio 4. The terms are then 5, 20, 80, 320, and the sum is

$$5 + 20 + 80 + 320 = 425.$$

Multiply each term and the sum by the ratio, 4, giving

$$\begin{array}{lr} & 2\not{0} + 8\not{0} + 32\not{0} + 1280 = 1700, \text{ 4 times the sum,} \\ \text{and subtract} & \underline{5 + 2\not{0} + 8\not{0} + 32\not{0} } = \underline{425}, \text{ 1 time the sum,} \\ \text{obtaining} & 1280 - 5 = 1275, \text{ 3 times the sum} \end{array}$$

$$\begin{array}{lrl} \text{But} & 1280 & = (\textit{last term}) \times (\textit{ratio}), \\ & 5 & = \textit{first term}, \\ \text{and} & 3 & = (\textit{ratio}) - 1. \end{array}$$

Expressing this result in words, we have

RULE V. *To find the sum of a geometrical progression, multiply the last term by the ratio, from the product subtract the first term, and divide the remainder by the ratio less* 1.

Using our previous abbreviations, and letting S represent the sum, this is

$$\text{V. } S = [(l \times r) - a] \div (r - 1).$$

This rule enables the sum to be calculated when the ratio and the extremes are known. As stated, it applies only to an increasing series. If the series is decreasing, the first term (a) is the greater extreme and the last term (l) the lesser, and these are to be interchanged in the symbolic form of the rule. In the divisor the ratio, a fraction, is then to be subtracted from 1.

Suppose that the first term, the ratio, and the number of terms are known, and the sum is required. The preceding rule requires the first *and last* terms to be known, the last term is found by Rule I. Combining Rule I and Rule V, therefore, we have

RULE VI. *To find the sum of a geometrical progression, raise the ratio to a power indicated by the number of terms, and subtract* 1 *from the result, then multiply the remainder by the first term, and divide this product by the ratio less* 1.

This is abbreviated as follows:

$$\text{VI. } S = [(r^n - 1) \times a] \div (r - 1).$$

This rule illustrates the manner in which different ones of the rules may be combined or used in succession to calculate different elements of a geometrical progression when certain others are known; as another illustration we shall give one more such rule.

Suppose the sum and the extremes are known: required, to find the ratio. This is done by reversing Rule V, which involves the same elements but with different ones known. By examination of V it is seen that the required rule is

RULE VII. *To find the ratio from the sum and extremes, subtract the first term from the sum, and also the last term from the sum, and divide the first remainder by the second.*

The abbreviated form of this rule is,

$$\text{VII. } r = (S - a) \div (S - l).$$

Many more such rules might be given, but those already given are sufficient to show the general principles of geometrical progression and the methods of making the calculations. Some of the rules will provide applications of logarithmic calculation, and others will be found of use in a later chapter in the simplified calculation of compound interest.

If a geometrical progression consists of only three terms, then, by the law of the series, the third is the same number of times the second as the second is times the first. In other words, the ratio of the first to the second is the same as that of the second to the third. Thus, if the terms of such a geometrical progression are 3, 6, 12, the extremes being 3 and 6 and the ratio 2, we have $\frac{3}{6} = \frac{1}{2}$ and also $\frac{6}{12} = \frac{1}{2}$. Then, $\frac{3}{6} = \frac{6}{12}$, or 3 : 6::6 : 12. This is a proportion in which 6 is the mean proportional of the extremes 3 and 12. This means that a proportion in which one term is a mean proportional of the other two is also a geometrical progression of three terms, the mean of the proportion being the second term of the progression of three terms, and the extremes being the same in both.

71. Infinite Geometric Progression and Repeating Decimal Fractions. In using the formula or rule VI above the value of the ratio r should be a number greater than 1. If r is less 1, that is, a common or decimal fraction, then the subtractions are reversed, that is, instead of $r - 1$ and $r^n - 1$ we must use $1 - r$ and $1 - r^n$.

If the number of terms n is very large and r is a fraction then the quantity r^n is very small, for when a fraction is multiplied by itself the result is a still smaller fraction. If the progression continues indefinitely (or "to infinity") then in comparison with the number 1 the result of raising the fraction r to indefinitely great powers is negligibly small, that is, r^n may be considered to be zero. Instead of the quantity $1 - r^n$ in the numerator of VI we then have simply 1, and the formula VI becomes simply

$$\text{VIII. } S = \frac{a}{1 - r}.$$

In words this is expressed by

RULE VIII. *When the ratio of a geometrical progression is less than one and the progression continues infinitely with no last term, the sum "to infinity" is simply the quotient of the first term divided by the difference between the number one and the fractional ratio.*

This rule finds an important use in converting certain unending (non-terminating) decimal fractions to common fractions, an operation which was postponed in article 34. This conversion can be performed exactly only for what are known as "repeating" decimal fractions, that is, fractions in which the same decimal figure or group of figures appears repeatedly with no other figures than these. Examples of repeating decimal fractions are 0.3333333333+, which is equal to $\frac{1}{3}$; 0.66666666+, which is equal to $\frac{2}{3}$; 0.21212121+ which is equal to $\frac{7}{33}$; 0.653535353+, which is equal to $\frac{647}{990}$. In this last example the first figure 0.6 is simply $\frac{6}{10}$ or $\frac{3}{5}$, and the repeating part is .0535353+.

Such repeating decimals are converted to common fractions by noting that each repeated group of figures is itself an exact decimal fraction which may be converted into a common fraction, and then the sum of these several fractions is a never-ending series. Furthermore, since each of these separate partial decimal fractions is made up of the same figures as every other and is simply shifted farther and farther to the right in succeeding parts, each part is equal to the preceding one divided by 10, 100, 1000, etc., depending on the number of places it is shifted. That is, the successive terms or parts of the fraction form an "infinite" geometric progression with the repeated group of decimal figures as the first term, and the "shifting" divisor or denominator as the ratio, and this ratio in the form of a fraction is less than 1.

In the repeating fraction .6666666+ above the first figure has the value .6 or $\frac{6}{10}$, the second is .06 or $\frac{6}{100}$ which is the first divided by 10 or multiplied by $\frac{1}{10}$, the third is .006 or $\frac{6}{1000}$ which is the second multiplied by $\frac{1}{10}$, etc. The fraction is therefore an infinite geometric progression with first term $a = \frac{6}{10}$ and fractional ratio $r = \frac{1}{10}$. The full value of the non-terminating repeating decimal fraction is therefore the sum "to infinity" of this series, and by rule or formula VIII this is

$$S = \frac{\frac{6}{10}}{1 - \frac{1}{10}} = \frac{\frac{6}{10}}{\frac{9}{10}} = \frac{6}{9} = \frac{2}{3}.$$

Similarly in the decimal fraction .6535353+ the repeating part is .0535353+ and when this is converted to a common fraction it is found to be $\frac{53}{990}$. The sum of this value and the first term or part $\frac{6}{10} = \frac{594}{990}$ is then found to be $\frac{647}{990}$, the full value of the complete repeating fraction.

The method of converting a repeating decimal to a common fraction is found by summarizing and formulating the above procedure into a concise rule, as follows:

RULE IX. *To express a non-terminating repeating decimal fraction as a common fraction, express the first distinct group of figures, which is repeated, as a separate decimal and write it as a numerator with the proper denominator* 10, 100, 1000, *etc. Use this fraction as the first term* (a) *of a geometrical progression.*

Note the number of places the next group of repeated figures is shifted to the right and write a fraction with 1 *as numerator and* 10, 100, 1000, *etc., for denominator (according as the shift is one, two, three, etc., places). This resulting fraction is to be used as the progression ratio* (r).

With the values of a and r just found calculate the sum S of the infinite G.P. by the rule or formula VIII. This value is the value of the original given repeating decimal.

72. Exercises and Problems.

1. The first term of an arithmetical progression is 5, the common difference 4, and the number of terms 8; what is the last term?

2. If the first term of an arithmetical progression is $\frac{2}{3}$, the common difference $\frac{3}{8}$, and the number of terms 20, what is the last term?

3. The extremes of an arithmetical progression are 3 and 15 and the number of terms is 7; what is the common difference?

4. The extremes of an arithmetical progression are 5 and 75 and the common difference is 5; what is the number of terms?

5. Find the sum of the arithmetical progression, the first term of which is 4, the common difference 6 and the last term 40.

6. A person wishes to discharge a debt in 11 annual payments such that the last payment shall be \$220, and each payment greater than the preceding by \$17. Find the first payment and the amount of the debt.

7. The first term of a geometrical progression is 6, the common ratio 4, and the number of terms 6. Find the last term.

8. The extremes of a geometrical progression are 2 and 512 and the number of terms is 5; what is the ratio?

9. The extremes of a geometrical progression are 2 and 1458 and the ratio is 3; how many terms are there?

10. The extremes of a geometrical progression are 3 and 384 and the ratio is 2. Find the sum of the series.

11. The first term of a geometrical progression is 7, the ratio 3, and there are 4 terms; what is the sum?

12. What is the sum of 5 terms of a geometrical progression with the first term 175 and the ratio 1.06?

13. The extremes of a geometrical progression being 2 and 686 and the sum 800, find the ratio.

14. A man traveled 13 days, each day traveling 5 miles more than on the preceding day, and his last day's journey was 80 miles. What was his first day's journey and how far did he travel altogether?

15. The distance between two towns is 360 miles. In how many days can it be covered by a man who covers 27 miles on the first day and 45 on the last, and increases his daily distance by the same amount each day?

16. A farmer pays a note of $1196 in 13 quarterly payments in such a way that each payment is greater by $12 than the preceding. What are his first and last payments?

17. A merchant pays a debt in yearly installments in such a way that each payment is three times the preceding; his first payment is $10 and his last is $7290. What is the amount of the debt and how many payments are made?

18. The tickets of a lottery bear odd numbers from 1 to 999 and the price of each in cents is twice its number. What are the proceeds of the lottery?

19. In some European countries (and in United States Army usage) the hours of the day are numbered 1 to 24. How many strokes will a clock make per day in striking those hours?

20. How many strokes does a common clock make in striking the hours of a whole day (no half hours)?

21. The rungs of a ladder diminish uniformly from 2 ft. 4 in. long at the bottom to 1 ft. 3 in. at the top. If there are 24 rungs (steps) in the ladder, what is the total length of wood in them?

22. A string is wound on an ordinary spinning top of conical shape, and there are 15 turns. The first and the last are respectively $\frac{1}{2}$ inch and $3\frac{1}{2}$ inches long. How long is the string?

23. A passing tramp offered to work for a farmer for 1¢ for a week, if the farmer would pay him 2¢ for the second week, 4¢ for the third week, 8¢ for the fourth week, and so on, doubling the pay each following week. The farmer, after calculating quickly that the first ten weeks would cost him only about ten dollars, hired the tramp for six months on those terms. Did he make a wise bargain, and if the bargain were carried out strictly, how much would it cost him in the entire period (26 weeks)?

24. A ball thrown vertically into the air falls and rebounds (bounces) 40 feet the first time, again falls and rebounds 16 feet the second time, and so on in the same ratio. What is the total distance it will have traveled after leaving the hand when it strikes the ground for the sixth time?

25. It is found by experiment that the number of bacteria in a sample of milk under certain conditions doubles every three hours. If conditions remain the same, what increase will there be in 24 hours?

26. Express the following non-terminating repeating decimal fractions as common fractions and reduce to lowest terms: 0.111111+; .3333333+; .21212121+; .6535353+; .898989+.

Part 3

MEASUREMENTS

Chapter 12

SYSTEMS OF COMMON MEASURES

73. Introduction. In the determination of extent, dimension, capacity or amount we are said to *measure* these quantities, or to make measurements. The process of measuring consists in comparing the thing to be measured with some standard of the same kind as the thing measured, the standard being one which is generally accepted and agreed upon. Such a standard of measurement is generally prescribed by custom or by law and is called a *unit* of measurement.

Common measurements are of several kinds, and each of these may be subdivided or classified into several similar kinds. The chief of the common measurements are those of

1. Length;
2. Surface, or area;
3. Volume, or capacity;
4. Weight, or force of gravity;
5. Time;
6. Angles, latitude and longitude;
7. Temperature, or intensity of heat;
8. Money, or value.

Each country, of course, has its own money system and it is assumed that the reader is familiar with that of his own country, so that money systems will not be treated in this book. Time, angles, latitude, longitude and temperature, being of a nature different from the first four kinds of measurement, will be treated in a later chapter. This chapter will, therefore, treat of measures of length, area, volume, capacity and weight.

In specifying a measurement, the unit of measure must be named and the number of such units given, as 3 feet, 4 pounds, etc. The numbers are those whose properties are treated in Part 2 of this book, but when combined with the name of the unit these are called concrete or *denominate numbers*. Thus, 3 is an abstract number and all the rules of numbers apply to it, but 3 feet is a denominate number and additional rules are used in calculations involving such numbers. If a denominate number contains more than one unit, as a length of 3 yards 2 feet and 7 inches, it is called a *compound* denominate number. Simple or compound denominate numbers are added, subtracted, multiplied, etc., according to the rules for ordinary abstract numbers, as we shall see later, but additional rules are needed in some of the conversions of these numbers.

As stated above, every country has its own system of money measure. In very early times every country had also its own distinct system of each kind of measurement, and sometimes there were several of each kind in one country. At the present time all the leading civilized countries except Russia, the United States and the British Empire use the same system of common measures, however, and in science that same system is used in all countries. That system is the *metric* system, sometimes called the French system because the French were the first to adopt it. The system in common use in the United States and the British Empire is called the *English system* because of the origin of most of the units in England.

In this book the English and metric systems will be treated. In this chapter we will study the measures and the units themselves, and in the next chapter, calculations which involve these systems of measures.

The ancients usually derived their smaller units of measure from some part of the human body. Thus, we find the *fathom* (distance between the outstretched hands); the *cubit* (length of the forearm); the *ell* (distance from elbow to end of the middle finger); the *foot* (length of the human foot); the *palm* (width of the hand). The Roman foot, as well as many other units, was divided into a dozen parts, called *unciae*, and from this comes the English word *inches*. The foot and the inch are the main English units of length.

For longer lengths, there was still less uniformity in ancient times. Thus, we find the Hebrew *half-day's journey;* the Chinese *lih* (distance a man's voice carries on a clear plain); the Greek *stadium* (probably the length of the race course); the *pace* (a long step); the *furlong* (length of a furrow); the *mille passus* (Latin, "thousand paces"); and from this the English *mile.*

In the year 1374 the *inch* was defined in English law as the length of "three barley grains, round and dry." Later other arbitrary measures were adopted by the government. The *yard* is now defined in the United States as $\frac{3600}{3937}$ of the *International Meter*, which will be described later. The *foot* is then taken as one third of the yard, and the *inch* is one twelfth of the foot.

It is curious to note what an important part the grain of wheat or barley has played in the establishment of units of weight in both ancient and modern times. The Greeks made four *grains* of barley equivalent to the carob seed or *keration*, and from this is derived the *karat* (or carat), the measure by which precious stones are weighed.

In England in 1266 the weight of the penny coin was defined as the weight of "32 wheat grains in the midst of the ear," and, again, about 1600, as "24 barley grains, dry and taken out of the middle of the ear." From this the *Troy grain* is defined as $\frac{1}{24}$ *pennyweight;* 20 pennyweights then make one *ounce* (French, "onze," eleven), and 12 ounces a *pound* (Latin, "pondus," a weight) of 5760 grains. From the earlier definition the pennyweight is 32 grains *avoir-du-pois* (French, "to have weight"); $13\frac{43}{64}$ pennyweights then made one ounce and 16 ounces one pound of 7000 grains. The avoirdupois system is now the common English measure of weight for all substances except jewelry, for which the Troy weight is used, and drugs in small quantity, for which the apothecaries' weight is used.

After the French Revolution at the end of the eighteenth century, France adopted the so-called *metric system* in which all the common measures are based on the *meter*, which is supposed to be one ten-millionth of the distance from the earth's equator to the pole. All other units are based on the meter and obtained from it in powers of 10; the metric system is therefore a *decimal* system. (The United States money system was based by Thomas Jefferson on the French decimal system of common measures.)

The metric system was made obligatory in France by law in 1837. It was made legal in the United States by Congress in 1866 and is now used by all scientists and all the scientific and technical departments of the United States government but, as stated before, it is not in general common use.

On account of its being a uniform decimal system, the metric system is the simplest system ever devised, and is the most suitable for an international system. So far, however, only the standard electrical units of measurement are internationally and uniformly derived from the metric system.

A. English Common Measure

74. Measures of Dimension. As measures of dimension, we will include those of *length* (linear measure), surface or *area* (square measure), and *volume* (cubic measure). The tables of measures will be stated without discussion and considered as self-explanatory. The methods of combining and reducing the measures as denominate numbers will be treated in the next chapter. The abbreviation is given for each unit.

Length

12 inches	= 1 foot (ft.)
3 feet	= 1 yard (yd.)
$5\frac{1}{2}$ yards	= 1 rod (rd.)
40 rods	= 1 furlong (fur.)
8 furlongs	= 1 mile (mi.)
1.15 miles	= 1 nautical mile (naut. mi.)
3 nautical mi.	= 1 league
7200 leagues	= earth's circumference at equator

Area

12^2 = 144 square inches (sq. in.)	= 1 square foot (sq. ft.)
9 sq. ft.	= 1 sq. yd.
$30\frac{1}{4}$ sq. yd.	= 1 sq. rd.
40 sq. rd.	= 1 rood (R.)
4 R.	= 160 sq. rd. = 1 acre (A.)
160 acres	= 1 quarter section
4 qtr. sect.	= 1 section = 1 sq. mi.
36 sq. mi.	= 1 township

Volume

12^3 = 1728 cubic inches (cu. in.)	= 1 cubic foot (cu. ft.)
27 cu. ft.	= 1 cu. yd.
16 cu. ft.	= 1 cord foot (cd. ft.)
8 cd. ft.	= 1 cord (of wood)
$24\frac{3}{4}$ cu. ft.	= 1 perch (of stone or masonry)

75. Measures of Capacity. Although, strictly speaking, volume or cubic measure is a measure of capacity or space content, the term "capacity" is usually (in measurement) understood to refer to the measures given in this article. These are the *apothecaries' liquid measure*, used in measuring medicines, standard *liquid measure*, used in measuring all other liquids, and *dry measure*, used in measuring grains, fruits, vegetables and the like.

LIQUID MEASURE

4 gills (gi.)	= 1 pint (pt.)
2 pints	= 1 quart (qt.)
4 quarts	= 1 gallon (gal.)
1 gallon	= 231 cubic inches
$31\frac{1}{2}$ gallons	= 1 barrel (bbl.)
2 barrels	= 1 hogshead (hhd.)
2 hogsheads	= 1 butt

DRY MEASURE

2 pints (pt.)	= 1 quart (qt.)
4 quarts	= 1 gallon (gal.)
2 gallons	= 1 peck (pk.)
4 pecks	= 1 bushel (bu.)
1 bushel	= 2150.42 cu. in.

APOTHECARIES' LIQUID MEASURE

60 minims (♏)	= 1 fluid drachm (f ʒ)
8 fluid drachms	= 1 fluid ounce (f ℥)
16 fluid ounces	= 1 pint (0)
8 pints	= 1 gallon (cong.)

The apothecaries' fluid measure is used only by pharmacists and physicians but is given here for comparison and reference.

76. Measures of Weight. The meaning of *weight* is often confused with that of *mass* or quantity of substance. Weight is not a property of any substance or object. The weight of any substance or object is simply the *force* or pull with which it is attracted toward the center of the earth by gravitation. The mass of an object is the quantity of substance or stuff of which it is made, and the weight is the mass multiplied by the acceleration (tendency to cause motion) of gravity.

The mass of an object remains always the same, but as the action of gravity is different at different places on and outside or inside of the earth, the weight of an object changes as it is carried from place to place. Thus, the weight of a person is different in a deep mine, at sea level, on a high mountain, near the equator, near either pole, etc. At any *one place*, however, the weight of a certain mass remains *always the same* and the weights of all objects furnish an exact method of comparing them and the masses of their substances.

There are three common English systems of weights: *Troy* weight, used in weighing jewelry and the like; *apothecaries'* weight, used in weighing small amounts of drugs for retail sale; and *avoirdupois* weight, used for all other ordinary purposes. Apothecaries' weight is used only by a specialized few but is given here for comparison and reference.

TROY WEIGHT

24 grains (gr.)	= 1 pennyweight (dwt.)
20 pennyweight	= 1 ounce (oz.)
12 ounces	= 1 pound (lb.)

APOTHECARIES' WEIGHT

20 grains	= 1 scruple (℈)
3 scruples	= 1 dram (ʒ)
8 drams	= 1 ounce (℥)
12 ounces	= 1 pound (℔.)

The Troy and apothecaries' grain, ounce and pound are the same.

Avoirdupois Weight

$437\frac{1}{2}$ grains (gr.)	= 1 ounce (oz.)
16 ounces	= 1 pound (lb.)
100 pounds	= 1 hundredweight (cwt.)
196 pounds flour	= 1 barrel (bbl.)
20 hundredweight	= 1 ton (T.) = 2000 lb.
2240 pounds	= 1 long ton (used in mining)

The dry measure *bushel* is usually given in pounds but the number of pounds per bushel is different for different commodities; it is also different in different states where it is prescribed by law, varying from about 48 to 80 pounds.

B. The Metric System of Measure

77. The Metric System. In the English system of linear measure the *foot* is the basic unit and the other units (inch, yard, etc.) are either multiples or subdivisions of this unit. The square and cubic measures are then simply the squares and cubes of the corresponding linear units. In the measure of capacity and weight, however (and, as we shall see later, in temperature), there is no relation to the linear unit, except that after the liquid gallon and the dry bushel were established, it was found that they contained respectively about 231 and 2150 cubic inches. There is not even such a relation as this between the weight and the other measures, except that it is found that one cubic foot of water weighs about $62\frac{1}{2}$ pounds. In all the measures of the English system, the multipliers and divisors used to obtain the various units from the foot, pound and gallon are greatly different and there is no regularity among them.

In the metric system all is different. To begin with, instead of two or three sets of measures for capacity, weight, etc., there is only *one* for each. These are without exception *all* obtained from one single unit, the *meter* of length. Furthermore, and this is a great simplification in measuring and calculating, all the various multiples and subdivisions are obtained from the primary unit simply by continually multiplying or dividing by 10, which in the decimal system amounts simply to shifting the decimal point. Because of this feature, all the tables of measure are similar, all fractional denominate numbers are decimal fractions, and measures of the same kind are converted from one unit to another simply by shifting the decimal point.

The metric unit of length is the international *meter*, which is one ten-millionth of the distance from the earth's equator to either pole.* It is found that the meter is about $1\frac{1}{12}$ English yard or more nearly $39\frac{3}{8}$ inches. The meter is divided into 10 parts, each being called a *deci*-meter (*deci-* being the Latin prefix indicating *one tenth*). The decimeter is divided into 10 parts, each of which is therefore $\frac{1}{100}$ meter and is called a *centi*-meter (Latin, *centi-*, "hundredth"). The centimeter is slightly less than $\frac{2}{5}$ inch, there being slightly more than $2\frac{1}{2}$ centimeters in one inch. The centimer is divided into 10 parts, each of which is therefore $\frac{1}{1000}$ meter and is called a *milli*-meter (Latin, *milli-*, "thousandth").

For measuring longer distances, the meter is multiplied by 10 to give the *deka*-meter (Greek, *deka*, "ten"); then 10 dekameters make one *hekto*-meter (Greek, *hekta*, "hundred"); 10 hektometers make one *kilo*-meter (Greek, *kilo*, "thousand"). The kilometer is about $\frac{5}{8}$ mile. The longest unit is the *myria*-meter, which is 10 kilometers, about 2 leagues.

The same system of multiples and subdivisions is used in all metric measures, the Latin prefixes indicating subdivisions and the Greek prefixes indicating multiples, as follows:

myria-	indicates	10,000
kilo-	"	1,000
hekto-	"	100
deka-	"	10
Main Unit	"	1
deci-	"	$\frac{1}{10}$
centi-	"	$\frac{1}{100}$
milli-	"	$\frac{1}{1000}$

This system of multiples and subdivisions makes the tables of the metric system easy to learn and remember.

The square and cubic measures are simply the squares and cubes of the linear measure.

The measures of capacity and weight are based directly on the linear measure, as will be explained for each in its place.

* The actual standard is a bar of platinum in Sébres, France, which was made to be one ten-millionth of the polar distance. Later measurements showed that the bar was not exactly this figure. In 1960 the meter was redefined as 1,650,763.73 times the wavelength of the orange spectral line of krypton-86.

78. Measures of Dimension. The three metric measures of dimension: *length* (linear measure), *area* (square measure), *volume* (cubic measure), are given here in the usual form of tables, with the standard abbreviations in parentheses.

LINEAR MEASURE

10 millimeters (mm.)	= 1 centimeter (cm.)
10 centimeters	= 1 decimeter (dm.)
10 decimeters	= 1 METER (m.)
10 meters	= 1 dekameter (Dm.)
10 dekameters	= 1 hektometer (Hm.)
10 hektometers	= 1 kilometer (Km.)
10 kilometers	= 1 myriameter (Mm.)

The myriameter is very little used, most long distances being expressed in kilometers, as long distances are expressed in miles in English measure instead of leagues.

For measuring very small lengths in scientific work, a unit called the *mikron* (or micron) is sometimes used. The mikron is $\frac{1}{1000}$ millimeter. The abbreviation is the Greek letter *mu* (μ).

SQUARE MEASURE

10^2 = 100 square millimeters ($mm.^2$)	= 1 square centimeter ($cm.^2$)
100 sq. cm.	= 1 sq. dm. ($dm.^2$)
100 sq. dm.	= 1 sq. m. ($m.^2$)
100 sq. m.	= 1 sq. Dm. ($Dm.^2$)
100 sq. Dm.	= 1 sq. Hm. ($Hm.^2$)
100 sq. Hm.	= 1 sq. Km. ($Km.^2$)
100 sq. Km.	= 1 sq. Mm. ($Mm.^2$)

The square myriameter is but little used. In measuring land the square dekameter is called an *are*, and the square hektometer is called the *hektare*. The hektare is about $2\frac{1}{2}$ acres, English measure.

CUBIC MEASURE

10^3 = 1000 cubic millimeters ($mm.^3$)	= 1 cubic centimeter (cc. or $cm.^3$)
1000 cu. cm.	= 1 cu. dm. ($cm.^3$)
1000 cu. dm.	= 1 cu. m. ($m.^3$)
1000 cu. m.	= 1 cu. Dm. ($Dm.^3$)
1000 cu. Dm.	= 1 cu. Hm. ($Hm.^3$)
1000 cu. Hm.	= 1 cu. Km. ($Km.^3$)
1000 cu. Km.	= 1 cu. Mm. ($Mm.^3$)

Very little occasion arises for the use of the cubic kilometer and the cubic myriameter.

79. Measure of Capacity. There is only one measure of capacity in the metric system, and it is used for all purposes of capacity measurement. It is based on the linear measure.

The standard unit of capacity is the *liter*, which is almost one cubic decimeter. The liter is equivalent to about one English liquid quart.

CAPACITY MEASURE

10 milliliters (ml.)	= 1 centiliter (cl.)
10 centiliters	= 1 deciliter (dl.)
10 deciliters	= 1 LITER (l.)
10 liters	= 1 dekaliter (Dl.)
10 dekaliters	= 1 hektoliter (Hl.)
10 hektoliters	= 1 kiloliter (Kl.)
10 kiloliters	= 1 myrialiter (Ml.)

The kiloliter and myrialiter are but little used. The hektoliter is equivalent to about $2\frac{4}{5}$ bushels, English dry measure.

80. Measure of Weight. As in the case of capacity, there is but one metric measure of weight. This also is based on the linear measure, and on an invariable property of pure water. It is used for all purposes of weighing.

Pure water at one and the same place weighs differently at different temperatures. At one particular temperature (given later) it is heaviest of all. This temperature is referred to as the temperature of *maximum density* of water. The weight of one *milliliter* of pure water at its maximum density is taken as the metric standard unit of weight and is called the *gram.** One English avoirdupois ounce is equivalent to about 28 grams.

WEIGHT MEASURE

10 milligrams (mg.)	= 1 centigram (cg.)
10 centigrams	= 1 decigram (dg.)
10 decigrams	= 1 GRAM (g.)
10 grams	= 1 dekagram (Dg.)
10 dekagrams	= 1 hektogram (Hg.)
10 hektograms	= 1 kilogram (Kg.)
10 kilograms	= 1 myriagram (Mg.)
10 myriagrams	= 1 quintal (Q.)
10 quintals	= 1 tonneau (T.)

The kilogram is equivalent to about $2\frac{1}{5}$ English avoirdupois pounds.

* The official standard is the Standard Kilogram, a platinum object kept at Sébres, France.

The *tonneau* (metric ton) is equivalent to about 2205 pounds, or very nearly one long ton. It is, as can be seen from the table, the weight of one cubic meter of pure water at its maximum density, that is, the content of a square tank one meter (about $3\frac{1}{4}$ feet) on the edge.

81. Conversion from One System to the Other. It is always of interest and in some kinds of work it is necessary, to express English measures in metric measures, and vice versa. For this purpose it is necessary to know the equivalents of the chief units in each system in terms of those of the other. For the benefit of those familiar with the English system, but perhaps not familiar with the metric system, a few such equivalents have already been given in various places for purposes of comparison.

In the table below are given the corresponding equivalents for con-

Unit	Equivalent		Unit	Equivalent	
	Exact	Approx.		Exact	Approx.
1 acre	.4047 hektare	$\frac{2}{5}$	1 mm.	.03937 in.	$\frac{1}{25}$
1 bushel	35.24 l.	$35\frac{1}{2}$	1 av. oz.	28.35 g.	$28\frac{1}{3}$
1 cm.	.3937 in.	$\frac{2}{5}$	1 Troy oz.	31.10 g.	31
1 $cm.^3$	.0610 cu. in.	$\frac{1}{17}$	1 peck	8.809 l.	$8\frac{4}{5}$
1 cu. ft.	.0283 $m.^3$	$\frac{1}{36}$	1 liq. pt.	.4732 l.	$\frac{1}{2}$
1 cu. in.	16.387 $cm.^3$	$16\frac{2}{5}$	1 pound	.4536 Kg.	$\frac{4}{9}$
1 $m.^3$	1.308 cu. yd.	$1\frac{1}{3}$	1 dry qt.	1.101 l.	$1\frac{1}{10}$
1 $m.^3$	35.31 cu. ft.	$35\frac{1}{3}$	1 liq. qt.	.9464 l.	1
1 cu. yd.	.7645 $m.^3$	$\frac{3}{4}$	1 $cm.^2$	.1550 sq. in.	$\frac{1}{6}$
1 foot	30.48 cm.	$30\frac{1}{2}$	1 sq. ft.	.0929 $m.^2$	$\frac{1}{11}$
1 gallon	3.785 l.	$3\frac{4}{5}$	1 sq. in.	6.452 $cm.^2$	$6\frac{1}{2}$
1 grain av.	.0648 g.	$\frac{1}{15}$	1 sq. mi.	259 hektares	260
1 gram	15.43 gr. av.	$15\frac{1}{2}$	1 $m.^2$	1.196 sq. yd.	$1\frac{1}{5}$
1 hektare	2.471 A.	$2\frac{1}{2}$	1 $m.^2$	10.76 sq. ft.	$10\frac{3}{4}$
1 inch	2.54 cm.	$2\frac{1}{2}$	1 sq. yd.	25.293 $m.^2$	$25\frac{1}{4}$
1 kilogram	2.205 lb.	$2\frac{1}{5}$	1 sq. yd.	.8361 $m.^2$	$\frac{4}{5}$
1 kilometer	.6215 mile	$\frac{5}{8}$	1 ton	.9072 met. ton	$\frac{9}{10}$
1 liter	.9081 dry qt.	$\frac{9}{10}$	1 long ton	1.017 met. ton	1
1 liter	1.057 liq. qt.	1	1 met. ton	1.102 tons	$1\frac{1}{10}$
1 meter	3.281 feet	$3\frac{1}{4}$	1 met. ton	.9842 long ton	1
1 mile	1.6093 Km.	$1\frac{3}{5}$	1 yard	.9144 meter	$\frac{9}{10}$

version of measures from either system into the other. Some of these equivalents, or *conversion factors*, as they are called, are exact. Others are not exact, that is, the conversions do not "come out even." In such cases they are given to the nearest exact decimal place and are correct *to as many figures* as shown. This table will be of use in the calculations of the next chapter. In this table the *names* of the units are given in alphabetical order without regard to the system to which they belong.

Chapter 13

CALCULATION WITH DENOMINATE NUMBERS

82. Reduction of Compound Denominate Numbers. We have seen that when a denominate number, say a length, is expressed in several units, as 3 yds. 2 ft. 6 in., it is a *compound* denominate number. From the table of English linear measure we know that 3 yds. = 9 ft., so that 3 yds. 2 ft. = 11 ft. Also 6 in. = $\frac{1}{2}$ ft., and so 3 yds. 2 ft. 6 in. = $11\frac{1}{2}$ ft. When the compound denominate number 3 yds. 2 ft. 6 in. is expressed as the *simple* denominate number $11\frac{1}{2}$ ft., the compound number is said to be *reduced* to feet.

The original compound number 3 yds. 2 ft. 6 in., or, as reduced, $11\frac{1}{2}$ ft., can also be reduced to inches. For 1 ft. = 12 in., and therefore $11\frac{1}{2}$ ft. = $11\frac{1}{2} \times 12 = 138$ in. It may also be expressed in yards, for 1 yd. = 36 in., and therefore 138 in. = $138 \div 36 = 3\frac{5}{6}$ yds. Similarly, a distance or length expressed as a compound number of any units may be reduced to any one unit, that is, a simple denominate number. In the same way any compound number, as an area, a volume, a capacity or a weight, may be reduced to a simple number in terms of any one unit.

As an example of such reduction, 1 ton 3 cwt. 5 lb. 4 oz. may be reduced to tons, cwt., lbs., or oz. To reduce it to lbs. we have 1 ton 3 cwt. = 23 cwt. = 2300 lb., and 5 lb. 4 oz. = $5\frac{1}{4}$ lb., so that 1 ton 3 cwt. 5 lb. 4 oz. = $2305\frac{1}{4}$ lb. To reduce 5 m. 3 dm. 6 cm. to cm. we have 5 m. = 50 dm., 5 m. 3 dm. = 53 dm. = 530 cm., and therefore 5 m. 3 dm. 6 cm. = 536 cm.

The illustrations given above show plainly that in order to reduce any compound number to smaller units it is only necessary to multiply each unit, beginning with the largest, by the number of the next smaller unit which it contains, add to the product the given number of that next smaller unit, and repeat these steps until the whole is expressed in terms of the desired smaller unit. This procedure can be explicitly stated as the following

RULE: (i) *Multiply the given number of the largest unit by the number which it contains of the next smaller unit, and to the product add the given number of that next smaller unit.*

(ii) *Multiply this sum by the number which this unit contains of the next smaller unit, and add the given number of that second smaller unit.*

(iii) *Repeat this process and continue until the given number of the smallest desired unit has been added. The final sum is the desired result.*

In order to express a compound number in larger units, proceed according to the following

RULE: (i) *Beginning with the next unit smaller than the desired larger unit, reduce this and all smaller units to the smallest given unit by the rule already given.*

(ii) *Using the result just obtained as numerator write a fraction whose denominator is the number of the smallest given units contained in the largest desired unit, and reduce this fraction to its lowest terms.*

(iii) *Write the mixed number whose fractional part is the fraction just obtained and whose integral part is the given number of the largest desired unit. This mixed number is the desired result.*

As an illustration of the first rule, we will reduce 4 bbl. 8 gal. 3 qt. 1 pt. 2 gi. to gills. We have first: 4 bbl. $= 4 \times 31\frac{1}{2} = 126$ gal. Add 8 gal., $126 + 8 = 134$ gal. Then 134 gal. $= 134 \times 4 = 536$ qt. Add 3 qt., $536 + 3 = 539$ qt. 539 qt. $= 539 \times 2 = 1078$ pt. Add 1 pt., $1078 + 1 = 1079$ pt. 1079 pt. $= 1079 \times 4 = 4316$ gi. Add 2 gi., $4316 + 2 = 4318$ gi. Therefore, 4 bbl. 8 gal. 3 qt. 1 pt. 2 gi. $=$ 4318 gi.

As an illustration of the second rule we will reduce the same compound number, 4 bbl. 8 gal. 3 qt. 1 pt. 2 gi. to barrels. Beginning with the next smaller unit, gallons, reduce the 8 gal. 3 qt. 1 pt. 2 gi. to gills by the first rule. This gives 286 gi. Also, from the table, 1 bbl. $=$ 1008 gi. Hence, 8 gal. 3 qt. 1 pt. 2 gi. $= 286$ gi. $= \frac{286}{1008}$ or $\frac{143}{504}$ bbl. Therefore, 4 bbl. 8 gal. 3 qt. 1 pt. 2 gi. $= 4\frac{143}{504}$ bbl.

In order to express a compound number in terms of some unit intermediate between its largest and smallest unit: first reduce the larger unit to the desired smaller unit by the first rule; next express all lower units as a fraction of that unit by the second rule; and finally write the mixed number composed of this fraction and the whole number of the desired units already found. This mixed number is the desired result.

As an illustration we will express 2 rd. 3 yd. 2 ft. 3 in. in yards. By the first rule 2 rd. 3 yd. = 14 yd. By the second rule 2 ft. 3 in. = $\frac{27}{36} = \frac{3}{4}$ yd. Hence, combining the two, 2 rd. 3 yd. 2 ft. 3 in. = $14\frac{3}{4}$ yd.

When these rules are applied to the metric system, the result is found by extremely simple calculations, since all the multipliers are 10 in linear, capacity and weight measure; 100 in square and 1000 in cubic measure. Consider the example already given: To reduce 5 m. 3 dm. 6 cm. to cm. According to the rule multiply by 10 each time and add the next units, giving 536 cm. Thus the first rule for reduction when applied to the metric system amounts simply to this for linear, capacity and weight measure: *Write in a single number all the figures giving the numbers of the various units, beginning with the highest unit at the left and ending with the lowest at the right.*

As illustrations, 3 Dl. 4 l. 5 dl. 3 cl. = 3453 cl.; 4 m. 2 dm. 7 cm. 8 mm. = 4278 mm.

To apply the second reduction rule to the metric system: *First express the given compound number in the lowest given unit as just explained, and then shift the decimal point one place to the left for each larger unit in the scale until the desired one is reached.*

Thus, to reduce 3 Kg. 4 Hg. 7 Dg. 6 g. 5 dg. to Dg.: First, write the compound number as 34765 dg. Then to get to Dg. shift the decimal point two places, giving 347.65 Dg. In Kg. this would be 3.4765 Kg.

The same method is used for square and cubic metric measure, except that the multipliers are 100 and 1000, respectively, instead of 10. Thus, 4 m.^{2} 6 dm.^{2} 43 cm.^{2} = 40643 cm.^{2}; 4 m.^{2} 73 dm.^{2} = 473 dm.^{3} = 4.73 m.^{2}; 15 m.^{3} 15 dm.^{3} 465 cm.^{3} = 15015465 cm.^{3} = 15015.465 dm.^{3}; etc.

83. Addition and Subtraction. Simple denominate numbers are added and subtracted according to the rules given in Chapter 2 for other numbers. Compound denominate numbers are added and subtracted in the same way, provided, of course, that only the same *units* or *kinds* of things be added or subtracted.

To add 2 yd. 1 ft. 4 in. and 3 yd. 1 ft. 5 in., write similar units in the same columns and add column by column:

yd.	ft.	in.
2	1	4
3	1	5
5	2	9

The sum is 5 yd. 2 ft. 9 in.

As many numbers as desired may be added, as usual.

Subtraction is handled in the same manner as addition. Thus

	T.	cwt.	lb.	oz.
From	7	17	53	12
To subtract	5	6	21	9
Remainder	2	11	32	3

If, in addition, the sum of any one column amounts to 1 or more of the next larger unit, that number is to be carried over to the next column, and the excess, in the units of the column being added, written at the foot of the column being added.

	bu.	pk.	qt.	pt.
	3	3	5	1
	4	2	7	1
	5	3	6	1
Adding directly,	12	8	18	3
Carrying over	14	2	3	1

As first added, in this example, the pt. total is 3, which is 1 qt. 1 pt. The 1 pt. is written down and the 1 qt. carried over to the next column, giving 19 qt., or 2 pk. 3 qt. The 3 qt. is written down and the 2 pk. carried over to the next column, giving 10 pk., or 2 bu. 2 pk. The 2 pk. is written down and the 2 bu. carried over to the next column, giving 14 bu. The sum is therefore 14 bu. 2 pk. 3 qt. 1 pt.

If, in subtraction, the minuend in any column is less than the subtrahend, one unit is borrowed from the minuend of the next larger unit column, reduced to the unit of the column being subtracted, and added to the minuend of that column. The subtraction is then carried out, and, as usual, when the next column is reached it must be remembered that one unit has been borrowed from the minuend.

	Km.	Hm.	Dm.	m.	dm.
From	7	6	5	4	3
To subtract	2	8	4	2	4
Remainder	4	8	1	1	9

Here 4 dm. is greater than 3 dm. but by borrowing and adding 1 m. = 10 dm. to the 3 dm., giving 13 dm., 4 dm. can be subtracted, giving a remainder 9 dm. Then, 2 m. is to be subtracted from 3 m. instead of 4 m., giving 1 m. The remaining columns are handled in the same manner, as becomes necessary.

This method applies to any measure in either the English or the metric system, but it is to be noted that in the metric system, where the multiplier is 10 (except in square and cubic measure) the subtraction, and also the addition, proceeds as in the case of ordinary columns of simple numbers. Thus, the last example is the same as subtracting 28424 from 76543, the remainder being 48119.

84. Multiplication and Division. In multiplication of compound denominate numbers the work proceeds as in multiplication of ordinary simple numbers, the partial products being found column by column as usual. If, after multiplication, the product for any one column amounts to 1 or more of the next larger unit, it is to be handled exactly as in addition. Thus, to multiply 2 cu. yd. 12 cu ft. 750 cu. in. by 5:

cu. yd.	cu. ft.	cu. in.
2	12	750
		5
10	60	3750
12	8	294

Direct multiplication gives 10 cu. yd. 60 cu. ft. 3750 cu. in. but 3750 cu. in. = 2 cu. ft. 294 cu. in., and when the 2 cu. ft. is carried over to the next column the total is 62 cu. ft. = 2 cu. yd. 8 cu. ft. The 2 cu. yd. is then carried over to the next column giving 12 cu. yd. The final product is, therefore, 12 cu. yd. 8 cu. ft. 294 cu. in.

Short division is usually suitable for division of compound denominate numbers, as illustrated in the following example: Divide 48 T. 18 cwt. 72 lb. 12 oz. by 6.

T.	cwt.	lb.	oz.
6)48	18	72	12
8	3	12	2

If in any of the separate column divisions there is a remainder, this remainder is reduced to the next lower unit and added to the next column to form the next dividend. If there is a remainder of the smallest unit, it is written over the divisor to form a fraction (of that unit) as in ordinary division. Thus, if the divisor is 5 in the last division we have:

T.	cwt.	lb.	oz.
5)48	18	72	12
9	15	74	$8\frac{4}{5}$

Here the remainder in tons is 3 which is 60 cwt. Adding 60 to the next column, 18, the dividend is 78 cwt. The remainder here is 3 cwt. which is 300 lb. Carried over this gives as the next dividend 372 lb. The remainder here is 2 lb. = 32 oz. Carrying this over the next dividend is 44 oz., giving a quotient of $8\frac{4}{5}$ oz.

In the metric system multiplication and division are particularly simple, amounting, according to the procedure described above, simply to decimal multiplication or division.

In any case where the divisor is greater than the number of any unit in the dividend, the quotient for that unit is zero and the units in the column are carried over to the next column as a remainder.

If desired, the entire dividend may be reduced to the lowest (or any other) unit by the method of article 82 and this simple number then divided by the divisor. Thus, to divide 5 yd. 2 ft. 10 in. by 20: Reducing the dividend, 5 yd. 2 ft. 10 in. = 214 in. The quotient is then 214 in. $\div$ 20 = $10\frac{14}{20} = 10\frac{7}{10}$ = 10.7 in. In any case a final fraction may be expressed as a decimal fraction if desired or necessary.

The discussion in this article so far applies to multiplication and division of denominate numbers by abstract numbers. If it is desired to multiply or divide one compound denominate number by another compound denominate number, the two in either case are first reduced to a single unit, and both to the same unit. The two numbers are then multiplied or divided as in the case of any two ordinary numbers. If, in multiplication of linear measure numbers, the multiplier and multiplicand are both of the same unit, the product is the square unit of the same kind. If three linear units are multiplied together or a square and a linear unit, the product is the cubic unit of the same kind. The products of other or different units possess no significance in terms of any particular unit.

All the methods explained in this and the preceding article apply to each kind of measure in both the English and the metric systems, it being simply necessary in any case to use the correct reduction numbers or factors in passing from one unit to the next larger or smaller.

85. Conversion Between Systems. In order to express a certain number of units of any particular kind in one system as a certain number of units in another system, it is only necessary to multiply the given number of units by the equivalent of 1 of those units in the desired unit of the other system. The equivalents of the most used units in both the English and the metric systems are given in the table in article 81. We will give a few illustrations of the use of this table in conversion of units.

To convert 9 inches into centimeters, multiply 9 by the centimeter equivalent of 1 inch, which is 2.54, obtaining $9 \times 2.54 = 22.86$ cm. Similarly, 40 Kg. $= 40 \times 2.205 = 88.2$ lb.; 10 gal. $= 10 \times 3.785 = 37.85$ l.; 12 m.$^2 = 12 \times 1.196 = 14.352$ sq. yd., or 12 m.$^2 = 12 \times 10.764 = 129.17$ sq. ft.; etc.

To convert a compound number from units of one system to those of another, first reduce it to a single unit by the rules already given and explained, and then multiply by the appropriate equivalent. After the conversion the result may be reduced to any desired unit in the new system.

For example, to convert 3 yd. 1 ft. 5 in. to centimeters: First, 3 yd. 1 ft. 5 in. = 125 in. Then, 125 in. $= 125 \times 2.54 = 317.5$ cm. This is also equivalent to 3175 mm., to 3.175 m., etc. Similarly, 7 Dl. 5 l. 4 dl. 6 cl. = 7546 cl. = 75.46 l., and 75.46 l. $= 75.46 \times 1.057 = 79.7$ liq. qt., approximately.

For approximate calculations or rapid mental conversions or estimates, the approximate equivalents given in article 81 as simple common fractions, may be used. Thus, 18 inches are very nearly $2\frac{1}{2} \times 18 = 45$ cm.; 14 m.^{2} are nearly equivalent to 154 sq. ft.; etc. These must not be used, however, for careful or exact work.

In using the decimal equivalents, the accuracy or approximation in the results is to be decided or determined as in any simple multiplication or division, as in Chapter 4.

86. Exercises.

In the following work, the tables of Chapter 12 are to be used, any necessary conversion between systems being accomplished by the use of the decimal equivalents in article 81 unless otherwise specified.

1. Change 100 miles to inches.
2. How many rods of fence will be required to enclose a square farm $1\frac{1}{4}$ miles on each side?
3. A grocer bought 12 bbl. of cider at \$3.60 a bbl., and after converting it into vinegar, he sold it at 15 cents a quart. What was his profit?
4. If a ship sails 150 leagues in a day, how many land miles are covered?
5. If 14 A. be sold from a field containing 50 A., how many sq. rd. remain unsold?
6. A man returning from South Africa wishes to sell 36 lb. 8 oz. of bort at 1 dollar 4.2 cents per pennyweight. What is the cash value?

7. A dealer plans to put 8 hhd. of tobacco, each weighing 9 cwt. 42 lb., into boxes containing 48 lb. each. How many boxes will he need?

8. A druggist bought 1 ℔. 10 oz. of quinine at $2.25 an ounce and sold it in 10-gr. doses at $12\frac{1}{2}$ cents each. What profit did he make?

9. Reduce nine tenths of a yard to feet and inches.

10. The distance from Boston to Buffalo is 438 miles; having traveled $\frac{3}{8}$ of this distance, how far yet remains to go, in miles and lower units?

11. Reduce 201458 inches to miles.

12. In a pile of wood containing 9000 cu. ft. how many cords and lower cubic units are there?

13. In order to measure up 846 bushels of corn with a dry quart measure, how many times must it be filled?

14. A coal dealer bought at the mine 175 long tons of coal at $6.75 per long ton and sold it retail at $19.50 per (short) ton. Calculate his profit.

15. Mt. Everest is very nearly 29062 feet high. Express this as a decimal number of miles.

16. Add 1 T. 17 cwt. 8 lb.; 5 cwt. 29 lb. 8 oz.; 1 cwt. 42 lb. 6 oz.; and 17 lb. 8 oz.

17. Add 6 yd. 2 ft.; 3 yd. 1 ft. 8 in.; 1 ft. $10\frac{1}{2}$ in.; 2 yd. 2 ft. $6\frac{1}{2}$ in.; 2 ft. 7 in.; and 2 yd. 5 in.

18. Add 31 bu. 2 pk.; $10\frac{7}{8}$ bu.; 5 bu. $6\frac{1}{2}$ qt.; 14 bu. $2\frac{3}{4}$ pk.; and $\frac{2}{3}$ pk.

19. From 18 lb. 5 oz. 4 dwt. 14 gr. subtract 10 lb. 6 oz. 10 dwt. 8 gr.

20. Subtract 15 rd. 10 ft. $3\frac{1}{4}$ in. from 26 rd. 11 ft. 3 in.

21. Multiply 5 mi. 4 fur. 18 rd. 15 ft. by 6.

22. Divide 111 bu. 2 pk. 4 qt. by 47.

23. Reduce: 125 Kg. to g.; 17 Km. to mm.; .573 m.^{2} to cm.^{2}; 27 m.^{3} to dm.^{3}

24. Convert: 25 Kg. to lb.; 300 Km. to mi.; 65 l. to liq. qt.; 20 liq. qt. to l.; 50 mi. to Km.

25. The Eiffel Tower in Paris is 300 m. high. How many feet and what decimal fraction of a mile is this?

26. The Empire State Building in New York is 1250 feet high. How many meters and what decimal fraction of a mile is this?

27. Cast copper is 8.8 times as heavy as an equal volume of water. What is the weight in Kg. and in lb. of 5 liters of copper.

28. A man takes 120 steps in walking 100 m. What is the average length of each step in cm. and in inches?

29. A train traveling 1 Km. per minute travels how many meters per second? (1 min. = 60 sec.)

30. Sound travels 332 m. per second. How long will it take to travel 1 Km. and also 1 mile?

Chapter 14

TIME, TEMPERATURE AND ANGLE MEASURE

87. Time Measure. All time measurement is based on the time required for the earth to turn once on its axis, that is, the time of one complete turn such as a wheel makes on its shaft or axle. This time for the earth's rotation is called the *day*. Since, on the average, half the day is dark and half is light at any one place, each of these half periods is distinguished from the other and divided into one dozen parts, and each of these twelve intervals is called an *hour*. The hour is, therefore, $\frac{1}{24}$ of a day.

Each hour is divided into five dozen smaller intervals called *minutes* (Latin, *minutas*, "small"), and each of these intervals is subdivided into five dozen smaller intervals, each being called a *second* (the "second" division).

The time required for the earth to travel once around the sun and return exactly to the starting point (any point in the circuit may be taken as starting point) is about $365\frac{1}{4}$ days, more nearly $365\frac{10463}{43200}$ days, which is 365 days 5 hours 48 minutes 46 seconds. This length of time is called the *year*. For civil and legal purposes the year is taken as 365 days.

When the year is taken as 365 days the time record or *calendar* (Latin, *kalendium*, "account book") is therefore almost 1 day behind the true time at the end of four years, so that 1 day is added to every fourth calendar year, thus giving every fourth year 366 days; this year is called a *leap year*.

Since the time which is not counted each year is not quite $\frac{1}{4}$ day, then in 4 years not quite a full day should be added. In 25 leap years the added time amounts to one day too much, so that the extra day is not added at the end of every 25th four-year period. Thus, the leap year is omitted every 100th year.

A period of 10 years is called a *decade* (Greek, *deka*, "ten"). Ten decades or 100 years make a *century* (Latin, *centuria*, from *centum*, "hundred").

A term sometimes used in speaking of very long periods of time in history is the *millennium* (Latin, *mille*, "thousand") which is ten centuries or 1000 years.

Summarizing the preceding definitions and relations in a table corresponding to those given in Chapter 12 for other measures we have the following table for time measure:

TIME MEASURE

60 seconds (sec.)	= 1 minute (min.)
60 minutes	= 1 hour (hr.)
24 hours	= 1 DAY (dy.)
365 days	= 1 civil year (yr.)
366 days	= 1 leap year
10 years	= 1 decade (dec.)
10 decades	= 1 century (C.)
10 centuries	= 1 millennium (mill.)

This system of time measure is used in all civilized countries and in both the English and metric systems.

The division of the year into a dozen *months* is a business and legal convenience, but as the division is irregular it is not, strictly speaking, a part of the table of time measure.

The century and the millennium are very little used in actual time calculations, but are convenient in stating the lengths of long periods of time.

Compound denominate numbers expressing time are reduced and used in calculation in the manner described for other denominate numbers in Chapter 13.

88. Historical Note. The Calendar. In ancient times it was thought that the seasons repeated themselves every 340 days, that is, the year consisted of 340 days. This estimate was not based on the rotation of the earth, as that was not known, but on the so-called "march of the seasons" and the apparent motions of the heavenly bodies. This time was first divided into shorter periods corresponding to the full set of changes of the moon and so were called *months*. Each of these periods seemed to be about 28 or 29 days and so there were 12 months in a year.

Later and closer observations of the heavenly bodies by the Babylonians indicated that the year consisted of about 360 days. This was for a long time considered the length of the year and was divided into 10 periods of 36 days each. Even though these did not coincide with the periods of the moon they were still called months. For the first several months of the year the ancient names were retained; these were the names of the gods and goddesses of various races of people, the sixth being named for the Roman high goddess Juno. Beginning with the seventh the remaining four months were named from the Latin words for the numbers which the Romans gave them, according to the following list:

September (*septum*, seven)
October (*octo*, eight)
November (*novem*, nine)
December (*decem*, ten)

By the time of the Roman emperor Julius Caesar it was known that the year consisted of $365\frac{1}{4}$ days. He, therefore, decreed that the legal year should consist of 365 days; that 6 hours in each year should be disregarded for three years and an entire day added to the fourth year, and that this day should be added to the second month.

Julius Caesar also decreed that the year should be divided into 11 months instead of 10, the new month being inserted between the old sixth and seventh months and named in honor of himself, *Julius*. From this we have the name *July*.

Not to be outdone, the emperor who succeeded Julius Caesar, Augustus Caesar, who ruled practically the entire civilized world at that time, decreed that the year should be divided into 12 months and that the new month should be named for himself, *Augustus*, from which we have *August*. This month was placed between the previous seventh and eighth months, following *Julius*. In this way the present list of months was completed and the words *septem*, *octo*, *novem*, *decem* now no longer have their original meaning in the calendar.

At present it seems that the list may have another month added, to bring the calendar month into agreement with the moon's 28-day period and give every month exactly the same length, consisting of exactly 4 weeks of 7 days each. The present average of the 12 months is about $4\frac{1}{3}$ weeks.

We have seen that the year actually consists of 365 dy. 5 hr. 48 min. 46 sec. instead of 365 dy. 6 hr. as was thought at the time of Caesar. By the year 1582 A.D. the error in the *Julian* calendar amounted to 10 days. To correct this error Pope Gregory decreed that 10 days should be stricken from the calendar, the day following October 3,

1582, to be called October 14th. This brought the spring equinox (equality of day and night) to March 21st. Most of the Catholic countries adopted this *Gregorian* calendar immediately.

Great Britain did not adopt the Gregorian calendar until 1752, at which time the Julian calendar was 11 days behind the true time. To correct this error the British Parliament decreed that 11 days should be stricken from the calendar, the day following September 2, 1752, being called September 14th. All the English-speaking colonies, later the English-speaking nations, adopted the same calendar, and it is now used by all the civilized countries, Russia being the last to adopt it. Russia adopted the Gregorian calendar in 1917, when Russian dates were 13 days behind those of the rest of the world. The change became effective at the beginning of February 1918, when the day following January 31st was called February 14th.

The centuries are now numbered from the beginning of the Christian era and marked A.D. (Latin, *anno domini*, "in the year of our Lord"). The years are numbered from the beginning of the century, the months from the beginning of the year, the days from the beginning of the month, and the hours from the beginning of the day, which is taken as midnight when the activities of the smallest number of people are affected by the change in date.

Using this scheme, 10:00 o'clock A.M., March 9, 1962, is the end of the 10th hour of the 9th day of the 3rd month of the 62nd year of the 20th century of the Christian era. The 7th decade of the 20th century began on January 1, 1961.

89. Temperature Measure. *Temperature* is the measure of the *intensity of heat*, not quantity of heat, nor either intensity or quantity of cold. "Cold" is the word used to indicate low heat intensity. The measurement of temperature is based on the intensity of heat required to freeze and to boil pure water under a pressure corresponding to that of the atmosphere at the average level of the sea. The instrument used in temperature measurement is called the *thermometer* (Greek, *thermos*, "heat"; Latin, *metrum*, "measure").

There are two common systems of temperature measure, the English or *Fahrenheit* system, named from the English physician who devised it, and the metric or *centigrade* system, named from the way the units are selected and numbered. The centigrade system is also sometimes called the *Celsius* system, from the name of the Swedish astronomer who invented it.

In the centigrade (or Celsius) system the freezing and boiling temperatures of water under standard conditions are taken as the standard reference points and the hundredth part of the difference between these temperatures is called one *degree* (*centi-grade* = "hundred degrees"). The freezing point of water is taken as zero degrees centigrade, written 0° C., and the boiling point of water is then 100° C.

Higher temperatures than the boiling point of water are then numbered accordingly. Thus, the melting temperature of iron is 1530° C.; the temperature of the electric arc is about 3600° C.; the surface temperature of the sun is about 6000° C.; that of the interior of the sun and other stars is estimated in millions of degrees. The average temperature of the human body is about 37° C.; the average "room temperature" is about 20° C.; water has its maximum density at 4° C.

Temperatures above 0° C. are called positive and indicated by a plus sign (+) when it is necessary to distinguish from temperatures below 0° C., which are called negative and indicated by a minus sign (−). Thus, the temperature of boiling water is +100° C.; the freezing temperature of mercury (quicksilver) is −38.9° C.; the temperature of liquid air is −192° C.; etc.

The sciences of physics and chemistry have shown that heat is a form of energy of motion and that the minutest particles (molecules) of all substances are in continual violent motion. This motion subsides and the volumes of ordinary substances decrease as their temperature decreases. In gases, the volume at 0° C. decreases by $\frac{1}{273}$ for each centigrade degree decrease in temperature. At this rate, all natural motion would cease and the volume of gases would disappear at −273° C. This hypothetical vanishing point temperature, than which there could be none lower, is called the *absolute zero*. Since all gases liquify on cooling before absolute zero could be reached, the "vanishing point" has no real meaning. However, absolute zero is the lower limit of the absolute temperature scale, which starts from −273° C. instead of 0° C. Absolute temperature is sometimes referred to as the *Kelvin* scale in honor of Lord Kelvin (Sir William Thomson) the great British scientist, who made a thorough mathematical investigation of the theory of temperature. In absolute temperature the freezing point of water is 273° K.; the boiling point of water is 373° K.; etc.

The centigrade scale of temperature is used wherever the metric system is used and in all scientific work, including that of United States Government departments.

The Fahrenheit system is used wherever the English system is used, being the system in common everyday use in the United States.

In the Fahrenheit system the difference between the freezing and the boiling points of water is divided into 180 degrees and the zero point is taken as 32° below freezing. This is denoted as 0° F. The freezing point is, therefore, +32° F. and the boiling temperature is +212° F. The average temperature of the human body is about +98° F.; the melting temperature of iron is 2786° F.; the temperature of liquid air is about −274° F.; the absolute zero is about −459° F.; etc.

The Fahrenheit and centigrade scales coincide at −40° on each.

90. Conversion of Temperature Measures. Since the interval from boiling to freezing temperatures for water is 100° C. and 180° F., then 1° C. = $\frac{9}{5}$° F., and 1° F. = $\frac{5}{9}$° C. Also, the centigrade zero is 32° F. Therefore, to express in degrees Fahrenheit a temperature which is given in degrees Centigrade, we have

RULE I. *Multiply degrees C. by $\frac{9}{5}$ and to the product add* 32. *The result is the same temperature in degrees F.*

Reversing this rule we have

RULE II. *Subtract* 32 *from degrees F. and multiply the remainder by $\frac{5}{9}$. The result is the same temperature in degrees C.*

If we let C represent the number of degrees centigrade and F the number of degrees Fahrenheit these two rules may be abbreviated and concisely expressed as follows:

$$\text{I.}\quad F = (\tfrac{9}{5} \times C) + 32°.$$

$$\text{II.}\quad C = \tfrac{5}{9} \times (F - 32°).$$

Thus a temperature of 45° C. is equal to $(\frac{9}{5} \times 45) + 32 = 81 + 32 = 113$° F. Similarly, a temperature of 95° F. is equal to $\frac{5}{9} \times (95 - 32) = \frac{5}{9} \times 63 = 35$° C.

In the case of negative temperatures (below zero) the words "add" and "subtract" in Rules I and II are to be interchanged.

91. Angles. When two straight lines meet at a point the figure that they form is called an *angle.* The size or measure of this angle is determined by the sharpness or bluntness of the "corner" between the two lines, and not by the length of either of the lines. In Figs. 1 and 2 the lines *AB* and *AC* are said to form the "angle *BAC*," and the lines *DE* and *DF* form the "angle *EDF*." The two lines which form an angle are called the sides of the angle.

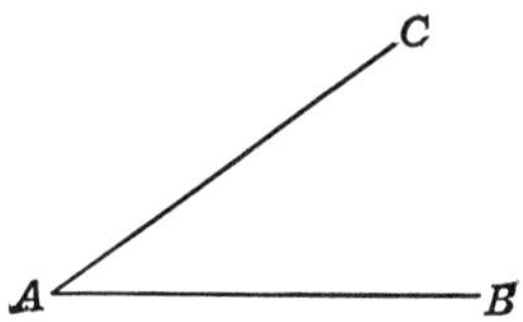

FIG. 1.

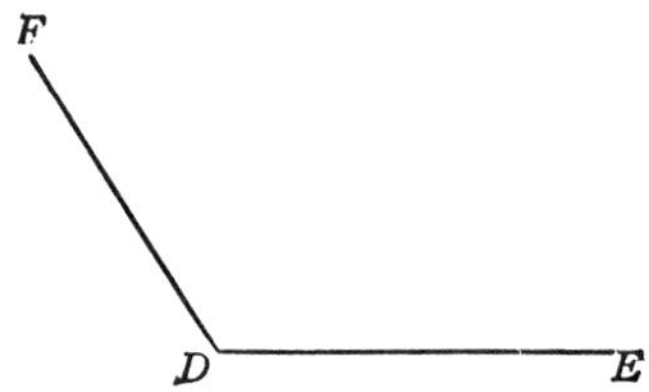

FIG. 2.

The angle *BAC* is sharper than the angle *EDF* and is said to be smaller than *EDF*. That is, in other words, the amount of separation, or *difference in direction* of the two lines *DE* and *DF* is greater than the difference in direction of the lines *AB* and *AC*, and angle *EDF* is greater than angle *BAC*. The size of an angle is given by stating the separation or *difference in direction* of its two sides. If two lines have the same direction their difference in direction is zero and they are said to form a zero angle.

If two lines of the same direction also pass through the same point they obviously coincide and amount to only one line. If two lines of the same direction do not coincide but are separate and distinct they are said to be *parallel*. From this viewpoint parallel lines remain always at the same distance apart and therefore never meet, no matter how far they extend.

FIG. 3.

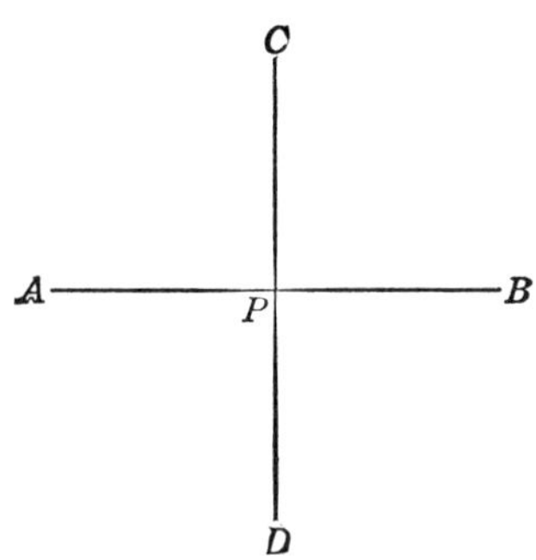

FIG. 4.

If the two sides of an angle form a "square corner" as in Fig. 3 the angle is called a *right angle*. The two sides are then said to be *perpendicular* to one another. In Fig. 4 also the two lines AB and CD meeting at the point P are perpendicular to one another, and form the four right angles BPC, CPA, APD, and DPB.

If an angle is less than a right angle it is called *acute* (sharp), while an angle greater than a right angle is called *obtuse* (dull). Thus, in Fig. 1 BAC is an acute angle and in Fig. 2 EDF is an obtuse angle.

Instead of thinking of an angle as being formed of two distinct lines, as AB and AC in Fig. 1, we may also think of it as being formed by the side AB turning about A as a pivot point until it reaches the position

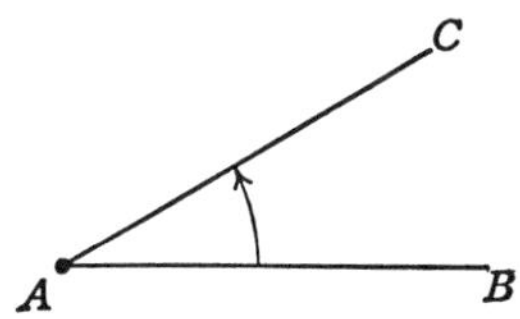

FIG. 5.

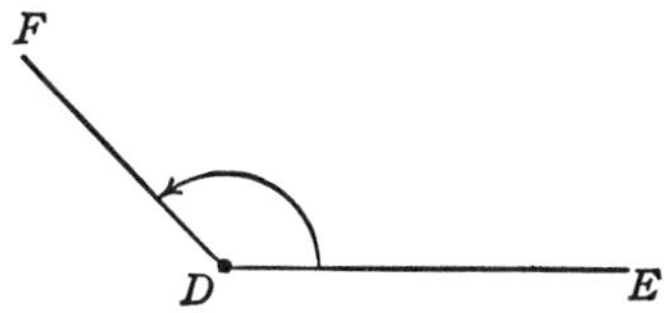

FIG. 6.

AC, as indicated by the curved arrow in Fig. 5. Similarly, as in Fig. 6, the line DE has formed or *generated* the angle EDF by turning about the point D from the *initial* position DE to the *terminal* position DF.

From this viewpoint, the four right angles of Fig. 4 can be said to be generated by the initial line PB as it turns about P to the successive positions PC, PA, PD, and finally again to the original position PB. When the rotating or generating line finally reaches its initial position after continuous turning in the same sense, it has swept through all possible directions and has turned through a circle. All possible *difference in direction* is, therefore, contained in a *circle*. That is to say, the total of all possible angles about a single point (the center) form a complete circle. Since the lengths of the sides of the angles, that is, the generating line in its successive positions, have nothing to do with the sizes of the angles, the size of the circle, or distance from its center, has nothing to do with the total of all angles in the circle about its center. The circle is, therefore, the natural reference unit or basis for all measurement of angles.

92. Angle Measure. It was thought by the ancient Babylonian astronomers and mathematicians that the year consisted of 360 days (see article 88). That is, as we now say, the earth passes over its circuit (circle) about the sun in 360 days, or, as the Babylonians said, the heavens turned about the earth in 360 days. In either case $\frac{1}{360}$ of the circuit or circle was thought to be completed in one day, and therefore $\frac{1}{360}$ of a circle was chosen as the unit of measure for angles. This unit of angle is called the *degree.*

On the basis of our first statement of the meaning of an angle, in connection with Fig. 1, if two lines are drawn from one point with such a *difference in direction* that 360 such lines as *PB*, *PC*, *PA*, etc., in Fig. 4 can be drawn, equally spaced, from the point then the angle between any two adjacent ones of the 360 lines is an angle of 1 degree. From the viewpoint of the rotating line, when it has turned from its initial position through $\frac{1}{360}$ of a circle it has turned through or generated an angle of one degree. Thus, an angle of one degree, written 1°, is an angle the *difference in direction* of whose sides is $\frac{1}{360}$ of a circle.

As seen in Fig. 4, four right angles constitute a whole circle of angles. Each is therefore one fourth or one *quadrant* of a circle, and a right angle is therefore equal to 90°.

For convenience in measurement and in calculation, the degree is divided into five dozen or 60 equal and more minute parts, each subdivision being called one *minute.* Minutes are indicated by the so-called *prime* (first) mark (′). Thus 45 minutes ($\frac{3}{4}$°) of angle is written 45′.

The minute is again subdivided into 60 equal parts called the second division or simply *seconds.* Seconds are indicated by the *second* mark (″). Thus 30 seconds ($\frac{1}{2}$′) of angle is written 30″. It was at one time proposed that seconds be divided into 60 *thirds* (‴), but fractions of a second are now generally indicated decimally. Thus $45\frac{1}{4}$″ is written 45.25″.

Summarizing the definitions and description of angle measure given in this article, it can be presented in the usual form of the table of angle measure:

ANGLE MEASURE

60 seconds (″)	=	1 minute (′)
60 minutes	=	1 DEGREE (°)
90 degrees	=	1 quadrant (quad.)
4 quadrants	=	1 circle (⊙)

This system of angle measure is the only one in general use in measurement and calculation, being used in both the English and metric systems of measure.

For certain theoretical purposes in higher mathematics, however, another system, derived also from the circle, is sometimes used. This system is studied in geometry and trigonometry.

The idea of angle and angle measure is useful in the applications of arithmetic to certain geometrical measurements, which are taken up in Chapter 16, and also in the study of latitude and longitude and their relation to time, which we take up in the following chapter.

93. Exercises and Problems.

1. If a youth's age is 18 yr. 14 dy., how many minutes old is he, allowing 4 leap years to have occurred in that time?
2. Stating your own age to the nearest whole day, or hour if it is known, find the number of seconds in that time, as well as the number of minutes and hours.
3. Reduce 90° 17′ 40″ to seconds of angle.
4. How many days, hours, minutes and seconds are there in 431,405 seconds?
5. If a man travels at the rate of 1′ of angle distance in 20 min. of time, how much time would he require to travel around the circle of the earth's equator?
6. How much time will a person gain in 36 yr. by rising 45 min. earlier and retiring 25 min. later every day, allowing for 9 leap years?
7. War between England and the United States began April 19, 1775, and peace was restored January 20, 1783. How long did the war continue?
8. What length of time elapsed from 16 minutes past 10 o'clock A.M., July 4, 1855, to 22 minutes before 8 o'clock, P.M., December 12, 1860?
9. On a certain day the Weather Bureau reports the temperature as 50° F. What is the temperature as read on the centigrade scale?
10. The surface temperature of the sun is about 6000° C. Express this temperature in degrees F.
11. Read the Fahrenheit thermometer as you read this and express the temperature in degrees centigrade.
12. Lead melts at 620.6° F. What is the centigrade melting point?
13. Pure water is heaviest at 4° C. What Fahrenheit temperature is this?
14. How many seconds of angle are there in 1°? In a circle?
15. How many minutes are there in a circle?
16. Reduce 15° 28′ 15″ to seconds.
17. Express 216,746″ in deg. min. sec.
18. What part of a circle is a 45° angle?
19. If there are 15 spokes in a wheel, what is the angle between any two adjacent spokes in degrees? How many minutes is this?
20. Through what angle do the hour and minute hand of a clock, respectively, turn (a) in one hour, and (b) in one minute of time?
21. Through what angle do the minute hand and the second hand of a watch, respectively, turn (a) in one minute, and (b) in one second of time?

Chapter 15

LATITUDE, LONGITUDE AND TIME

94. Latitude and Longitude. In order to furnish a simple method of locating positions on the earth, two sets of lines are drawn on maps of the earth and *imagined* as drawn on the earth itself to be used as standard reference lines. These sets of lines are shown on the globes of Fig. 9. We shall hereafter refer to these lines as drawn on the earth; actually, of course, they are not.

The set of lines running round the globe (earth) parallel to the equator *WE* (Fig. 9) are called lines or *parallels of latitude*, and the set of lines crossing these and passing through the poles are called *meridians of longitude*. These lines are numbered in a manner which is explained below and any place on the earth is located by stating the numbers of the latitude and longitude lines which pass through it; that is, by giving the *latitude and longitude* of the place.

The earth is a globe, almost an exactly round ball. The imaginary straight line passing through its center and about or around which it turns as a "shaft" or "axle" is called its *axis*, and the two ends of this line on the surface of the earth are called the earth's *poles*. The end which points toward a certain star in the heavens ("Polaris") is called the *north pole*, and the other is called the *south pole*. These are marked *N* and *S*, respectively, on globes and maps. A line imagined drawn around the earth midway between the poles is called the *equator*, as it divides the earth's surface into two equal parts. The equator *WE* and poles *N*, *S* and the parallels of latitude, parallel to the equator, are represented in Fig. 9(*a*).

The manner of locating these lines is shown in Fig. 7. Since the earth is round, any line drawn around it, passing through the poles and making right angles with the equator, is a circle. Such a circle is called a *meridian* circle and is marked off in degrees of angle as shown in Fig. 7, and explained in article 92. The parallels of latitude are drawn through these points of division, running east and west. Since the equator and the poles divide the meridian circle into quadrants, there are 90 degrees between the equator and each pole along any meridian circle. The parallels of latitude are then numbered from the equator to each pole, northward and southward, beginning with 0° at the equator and ending with 90° at each pole. These lines are shown for every 10° on the meridian circle in Fig. 7 and the

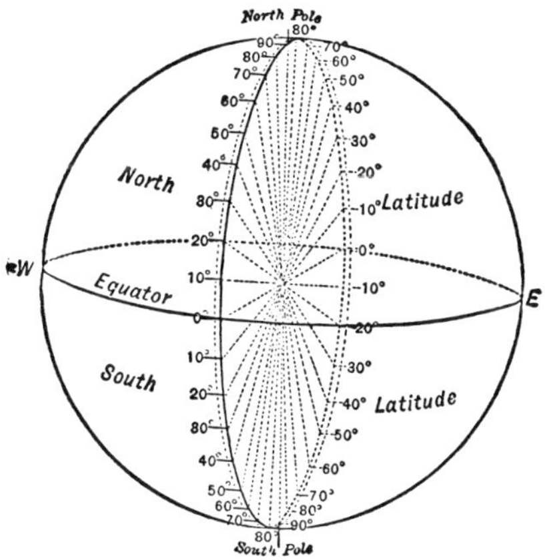

FIG. 7.

lines from the center making these 10° angles are shown dotted. The degrees of latitude are further divided as usual into minutes and seconds. In a small diagram each degree, minute and second cannot be shown.

Places between the equator and the north pole are said to be in *north latitude* and places between the equator and the south pole in *south latitude.* In order to tell exactly how far north or south a place is from the equator, it is only necessary to state its north or south latitude in degrees, or, to be very precise, in degrees, minutes and seconds. For example, the city of New York, N. Y., is about 41° N. A reference marker placed at the Museum of Natural History in New York City is at exactly 40° 46′ 47.17″ N.

For the meridian lines of longitude the equator is the circle on which the divisions are marked off in degrees, or degrees, minutes and seconds, as shown in Fig. 8. The meridian lines through the division points all pass through the poles and, therefore, draw closer and closer together as the poles are approached. There is no natural dividing line for longitude, as the equator is for latitude, so by common consent of the civilized world the meridian line passing through the British Royal Astronomical Observatory at Greenwich, England, near London, is taken as the reference or *prime meridian*, as shown in Fig. 8.

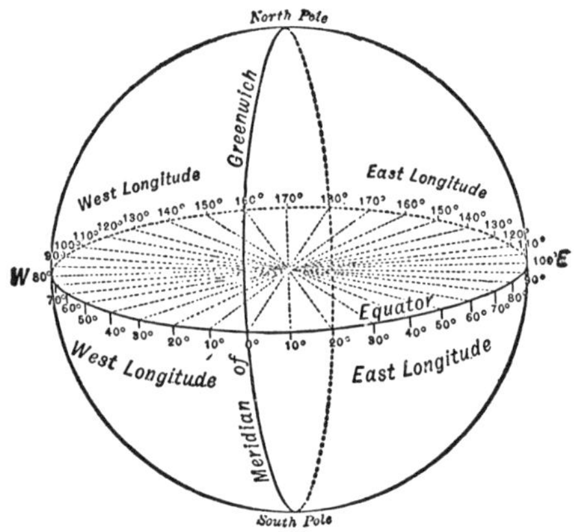

FIG. 8.

The Greenwich prime meridian divides the equator into two semicircles of 180° each and longitude is measured eastward and westward from this meridian, beginning at 0° where it crosses the equator south of Greenwich and continuing to 180° at the point where it crosses the equator on the opposite side of the earth.

Any place in the western half or hemisphere thus marked off, whether on the equator or not, is said to be in *west longitude* and any place in the eastern hemisphere is in *east longitude*. In order to tell how far east or west a place is from the prime meridian it is only necessary to state its east or west longitude in degrees, or to be precise, in degrees, minutes and seconds. For example, the city of New York is about 74° W. The reference marker in New York City, mentioned above, is at 73° 58′ 41.00″ W.

In Fig. 9 is shown the complete scheme of latitude and longitude. In both diagrams the *N* and *S* poles and the *W* and *E* directions are shown. The equator *WE* and the prime meridian are shown in heavy black lines and every 10th degree of both latitude and longitude is shown and marked. Fig. 9(*a*) is a broadside view showing both poles with the longitude marked along the equator and the latitude marked along the 70° E. meridian. In Fig. 9(b) the globe is turned so that parallels of latitude may be seen running entirely around the earth, and only the north pole appears. In this diagram longitude is marked on the equator and latitude on the prime meridian.

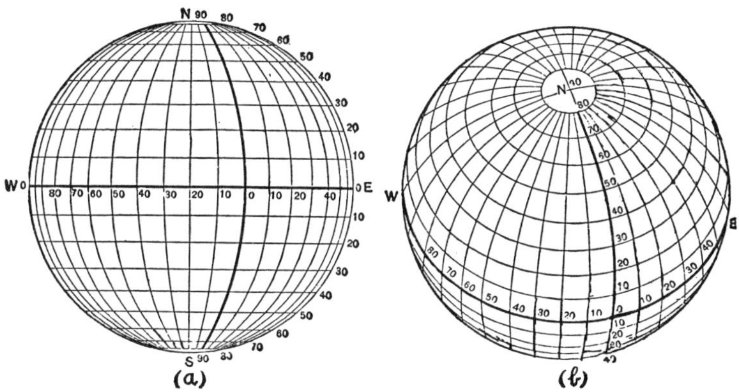

FIG. 9.

From Fig. 9 it is now clear that any point on the earth can be exactly located by giving the numbers of the two lines, of latitude and longitude, which pass through it and stating whether it is north (N) or south (S) of the equator and east (E) or west (W) of the prime meridian. The letters N, S and E, W are generally used for this purpose as indicated. Thus the marker in New York City already mentioned is completely located by saying that it is at 40° 46′ 47.17″ N., 73° 58′ 41.00″ W. (In place of these letters, however, the plus sign (+) is sometimes used for N and W and the minus sign (−) for S and E.)

Since the meridians of longitude are not everywhere equally spaced, they cannot be shown exactly on an ordinary flat map. The same difficulty is met here that would be met in trying to spread an orange peel out flat on a sheet of paper without distorting it. Map makers have certain methods of showing different parts of the earth's surface in different sizes or *scales* on maps. On some maps, therefore, the parallels and meridians appear as straight lines and on others as curved lines. From these the latitude and longitude of any place on the map can be determined at once, and from this information distances between different places can be determined from the relation of degrees of latitude and longitude on the earth's circles to the distance around its circumference. This relation will now be explained.

95. Latitude, Longitude and Distance. The earth is almost exactly round and the greatest distance around it (the *circumference*) is very nearly exactly 24,912 land miles, or 21,600 nautical miles. Since the circumference consists of 360° of angle, then 1° on the circumference circle is equivalent to $24{,}912 \div 360 = 69.2$ or $69\frac{1}{5}$ land miles.

It is this relation between distance and angle measure on the earth which makes the use of the nautical mile so convenient in navigation. By choosing the mile so that 1′ angle = 1 mi. distance there will be 60 nautical miles to 1° and angles and distance will then be easily converted, one to the other.

But in standard English miles there are, as seen above, 69.2 miles per degree. Then, since the parallels of latitude are everywhere at the same distance apart, one degree of latitude is always and everywhere very nearly 69.2 miles. One minute of latitude is therefore $69.2 \div 60 = 1.15$ land miles (1 nautical mile); and 1″ lat. $= 1.15 \div 60 = .0192$ mi.

Since the meridians of longitude are not everywhere at the same distance apart for the same angle difference, being farthest apart at the equator and meeting at the poles, the distance corresponding to one degree of longitude is greatest at the equator, zero at the poles, and different along every parallel of latitude crossing the meridians. Along the equator, 1° of longitude is very nearly $24{,}912 \div 360 = 69.2$ miles, and along every parallel of latitude farther and farther from the equator it is less and less. Along every 10th parallel of latitude the length in miles of each degree of longitude is given in the following table. For parallels intermediate between those given, the corresponding approximate distances are proportionately intermediate between those given. That is, for each additional degree between any two consecutive given angles, there is to be subtracted one tenth of the difference in distance corresponding to those two angles.

As an example, between 20° and 30° the distance difference is 6 miles for the 10 degrees, and therefore for each degree increase the distance decreases .6 mile. For, say, 24°, therefore, the distance is $65 - (4 \times .6) = 65 - 2.4 = 62.6$ miles. This means that along the 24th parallel of latitude the distance corresponding to one degree of longitude is about 62.6 miles.

Parallel of Latitude..	0°	10°	20°	30°	40°	50°	60°	70°	80°	90°
Length in Miles of 1° Longitude......	69.2	67.9	65.0	59.0	52.3	44.4	34.5	23.6	11.9	0

The preceding paragraphs explain how distances along the meridians of longitude and along the parallels of latitude are determined. These methods do not apply to distances along lines in other directions, and there are no reference lines on the earth or the maps for such other directions. The distance along any such other direction can be found, however, by special calculations when both the latitude and the longitude of each end of a line in such a direction are known. These special calculations are studied in trigonometry: plane trigonometry explains the calculations of short distances, and spherical trigonometry the calculations of long distances such as are necessary in sailing the seas in ships or crossing the land and sea in airplanes, and in drawing maps of large portions of the earth's surface.

96. Longitude and Time. The earth turns on its axis once in 24 hours and in one complete rotation it turns through its full 360° of longitude. In each hour, therefore, it turns through $360 \div 24 = 15$ degrees of longitude. This relation is expressed by saying that one hour of time is *equivalent* to 15 degrees of longitude. From this, the complete set of relations between longitude and time can be expressed in terms of hours, degrees, minutes and seconds. In this way navigators and astronomers use longitude and time interchangeably and carry out calculations of position and distance on the earth by means of the time. The method of doing this is indicated in article 97. Here we wish to show the general relation between longitude and time of day or night.

This relation is indicated by Fig. 10. Here the directions of the sun's rays is shown by the straight lines with arrow heads, and the earth is shown in the position which it occupies at the time of the equinoxes (when day and night are each 12 hours). The white part of the globe is in daylight and the shaded part in darkness. As the earth turns toward the east, as indicated by the curved arrow on the equator, the sunlight or daylight moves toward the west and the sun "rises" at any place (or that place rotates into sunlight) at 6:00 A.M. of the clock, and "sets" (as the place rotates into darkness) at 6:00 P.M. Then at every 15th meridian the sun rises (and sets) one hour later, so that at each 15th meridian the time is one hour later, since each place sets its clock as the sun rises.

At the moment when the midday rays shine on a certain meridian, it is noon for that meridian, and as we go east from that meridian, it is one hour later for every 15 degrees, or 2 hours for every 30°, as shown in the diagram, Fig. 10. As we go west, it is 2 hours earlier for every 30°, until at 180° from the noon meridian it is midnight.

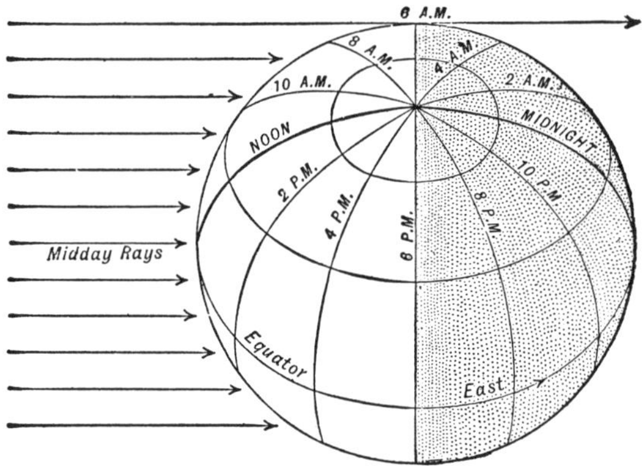

FIG. 10.

A slightly different view is shown in Fig. 11. This is a map of the northern half of the earth as it would appear from a point directly over the north pole, at the time of the autumn equinox. When it is noon of September 20th at the 90° W., or +90° meridian, it is midnight of September 20–21 at 90° E., or −90°, exactly opposite.

97. Longitude and Time Calculations. From the relations explained in the preceding article, time can be expressed in terms of longitude or longitude in terms of time, and from the difference in time of two places, their difference in longitude is known, and vice versa. From these relations, the corresponding distance can also easily be found from the known relation of longitude to distance when the places are at the same latitude, as given in article 95. The longitude-time relation is based on the fact that the earth actually rotates, or the sun *apparently* revolves, through 360 degrees in 24 hours.

From this, as we have seen, 1 hour of time is *equivalent* to $360 \div 24 = 15°$ of longitude. Then, 1 minute of time is equivalent to $15° \div 60 = 15'$ of longitude, and 1 second of time to $15' \div 60 = 15''$ of longitude.

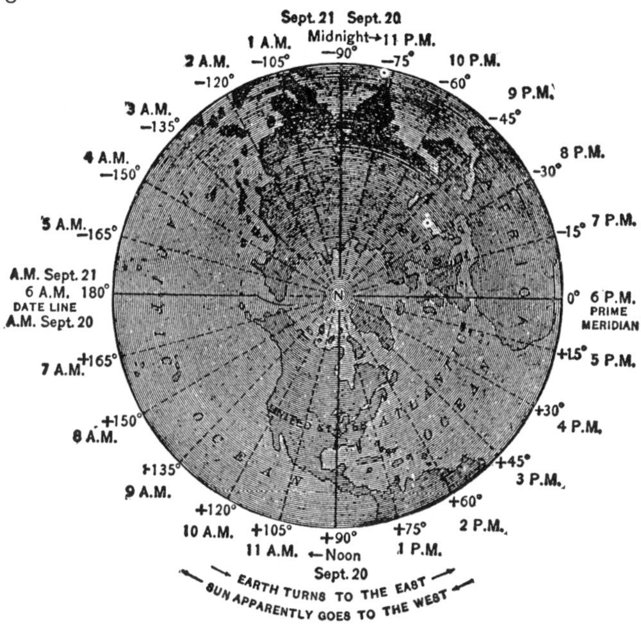

FIG. 11.

Similarly, since 360° of longitude is equivalent to 24 hours of time, 1° longitude is equivalent to 24 hr. $\div 360 = \frac{1}{15}$ hr. = 4 min. of time; 1′ longitude is equivalent to 4 min. $\div 60 = \frac{1}{15}$ min. = 4 sec. of time; and 1″ longitude to 4 sec. $\div 60 = \frac{1}{15}$ sec. of time.

The complete set of these relations is summarized in the following table:

LONGITUDE–TIME MEASURE

I			II	
Long.	*Time*		*Time*	*Long.*
360°	= 24 hr.		24 hr.	= 360°
1°	= $\frac{1}{15}$ hr.	= 4 min.	1 hr.	= 15°
1′	= $\frac{1}{15}$ min.	= 4 sec.	1 min.	= 15′ = $\frac{1}{4}$°
1″	= $\frac{1}{15}$ sec.		1 sec.	= 15″ = $\frac{1}{4}$′

From these tables it is now readily seen that difference of longitude is expressed as difference of time by

RULE I. *Divide the difference in longitude, as expressed in degrees, minutes and seconds by* 15, *by the rule for division of compound numbers; the quotient is the corresponding difference of time in hours, minutes and seconds.*

Reversing this rule, we have for converting time difference into longitude difference,

RULE II. *Multiply the time difference, as expressed in hours, minutes and seconds, by* 15, *by the rule for multiplication of compound numbers; the product is the longitude difference in degrees, minutes and seconds.*

The difference in longitude between two places is found by subtraction if both are east or both west, and by adding if one is east and one west, and if the result is greater than 180° it must be subtracted from 360° to obtain the true difference.

Since the sun appears to move from east to west, then when it is 12:00 o'clock at one place, it is past 12 at all places east and before 12 at all places west of that place. Hence, if the time difference between two places be *subtracted* from the time of the *easterly* place, the result will be the time of the westerly place; and if the difference be *added* to the time of the *westerly* place the result will be the time of the easterly place. From these relations, the rules already given, and the distance relations given in article 95, we get immediately the following rules for places *on the same parallel of latitude:*

RULE III. (i) *Express the longitude difference in degrees and decimal fractions of a degree, by the reduction rule for compound numbers.*

(ii) *Multiply this result by the distance found from the table of article* 95 *for the parallel of latitude on which the places are located. The result is the distance between the places.*

If the time difference is given, to find the distance it is first expressed as longitude difference by Rule II and Rule III is then used. To find time difference from distance *on the same parallel*, simply reverse Rule III; that is, divide by the distance number.

98. International Time and the Date Line. It has been agreed among the nations of the earth that all places on any one meridian of longitude shall have the same clock time, based on the longitude-time relations. Thus, for example, when it is noon at +90° (W.) as in Fig. 11, then the time shown for any meridian will be the time for all the places along that meridian, in both the northern and southern hemispheres. This time is called *International Standard Time.*

Of course, the actual correct time changes continuously and even though two places are only about 69 miles or 1° apart, east and west, on the equator (about 53 miles at the latitude of New York) there will be 4 minutes difference in their actual time. For convenience, however, the clocks are only changed when the difference is 1 hour, as shown in Fig. 11. Then all places in a region extending about $7\frac{1}{2}°$ on either side of every 15th meridian will have the same time.

In spite of the general agreement, there are some localities which still adhere to some early custom and do not have exactly standard time, but, in the main, standard time is used by most of the world.

Referring to Fig. 11, when it is noon of a certain day at the prime meridian then it is midnight between that day and the next at the 180° meridian; the time at other meridians is shown on the map of Fig. 12. The 180° meridian passes through the Pacific Ocean far

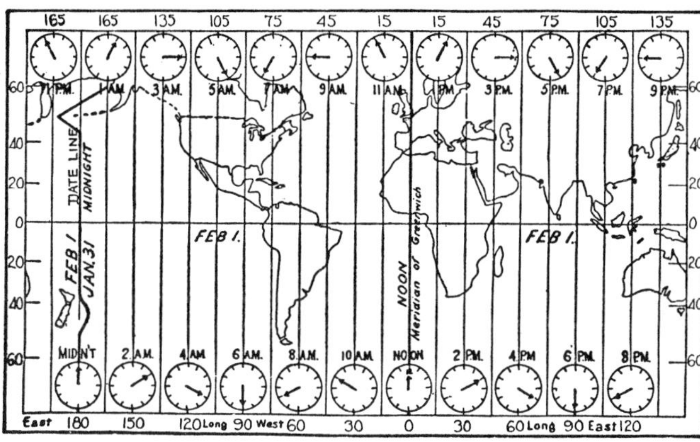

FIG. 12.

from any land except a few small islands and is called the *International Date Line*. The significance of this line will be made clear by the following considerations.

Suppose two men to start from the prime meridian at noon of a Monday and travel, the one eastward and the other westward, as fast as the earth rotates. The westward traveler keeps always directly beneath the sun and it seems to him to be still Monday when he reaches his starting point 24 hours later, although it is actually Tuesday noon. He has, therefore, *lost* a day in his time record.

The eastward traveler moves over the earth as fast as the earth itself turns and in the same direction, and therefore he moves away from the sun twice as fast as the prime meridian does. After 12 hours he reaches the 180° meridian, but 12 hours rotation of the earth has carried this meridian beneath the sun so this traveler reaches it at noon. In 24 hours he reaches his starting point, but 24 hours' rotation has brought this meridian under the sun again so this traveler reaches it at noon, to him the *second* noon after his start. He, therefore, supposes it to be Wednesday noon, when actually it is Tuesday noon. He has, therefore, *gained* a day in his time record.

To correct such errors, travelers and mariners must somewhere *add* a day to their records when traveling westward and *subtract* a day when traveling eastward. The meridian where this is done is that of 180°, the date line. A mariner sailing east and arriving at the date line on Monday would change his date to Sunday on crossing it, and thus would have two Mondays in the same week. But if sailing west and reaching the line on Sunday, he would change it to Monday on crossing and thus have no Sunday at all.

Let us suppose that any day, say January 31st, is just beginning at midnight on the date line. As midnight moves westward around the earth each place will begin the new day. By the time it is midnight at London (Greenwich) and the day just beginning, it will be noon at the date line. When the day begins at San Francisco it will be about 8:00 P.M. at the line. And just as January 31st is ending at any place east of the date line February 1st is beginning at any place west of the line. Thus the whole world except those places crossed by the date line have the same *name* and *date*, or number, for the same day.

The map of Fig. 12 represents the earth when it is noon February 1st at Greenwich on the prime meridian. It is, therefore, one hour earlier for each 15° east, as shown. Hence at 180° west it is midnight of January 31st, and at 180° east it is midnight of February 1st. That is, at the date line the date changes from January 31st to February 2nd instead of February 1st, in going west; and changes from February 1st to January 31st in going east; thus gaining a day in one direction and losing a day in the other.

As seen in the maps of Figs. 11 and 12, the date line is not exactly straight throughout its length. This is agreed upon in order to avoid confusion in the few small islands the line would cross and to allow them to have the same date as the nearest continent. The prime meridian was chosen to pass through Greenwich so that the date line 180° away would pass through the Pacific Ocean and near no important land.

99. National Standard Time. On examining the time maps in Figs. 11 and 12 it is seen that the west meridians of 75°, 90°, 105°, and 120° cross the United States. This country will, therefore, have four different standard times with a total difference of three hours. These would follow the lines shown in Figs. 11 and 12 if the standard time, with hour differences for each 15° were strictly adhered to, but for the convenience of the railroads and certain large cities these lines are not followed exactly. The actual dividing lines are more nearly those shown in Fig. 13.

The sections of the country using the same standard time are shown in Fig. 13 by heavy lines, and each of these sections is called a *time belt.* These belts are called the Eastern, Central, Mountain and Pacific time belts and the time in these belts is referred to as *Eastern, Central, Mountain, Pacific Standard Time,* respectively.

The Eastern belt extends about $7\frac{1}{2}°$ east and west of the 75° meridian and all places in this belt use the international standard time of the 75° meridian. Similarly, the remaining belts extend about $7\frac{1}{2}°$ on both sides of the 90°, 105°, and 120° meridians and each uses the international standard time of these three respective meridians. Most of Canada and Mexico use the same belts as the United States, and similarly other countries which extend very far east and west divide their territory into time belts. When any country thus has several time belts the time of the different belts is referred to as their *national standard time.*

100. Problems in Latitude, Longitude and Time. The subject matter of this chapter requires for its detailed study a knowledge of geometry and of plane and spherical trigonometry, and finds its practical applications and uses in the technical subjects of surveying, mathematical geog·aphy (or geodesy), map-making, and navigation (both marine and aerial). This same subject matter enters so closely into matters of everyday life, travel and world news, however, that it is important to be able to solve simple problems involving position, distance and time in various parts of the world. We solve and ex-

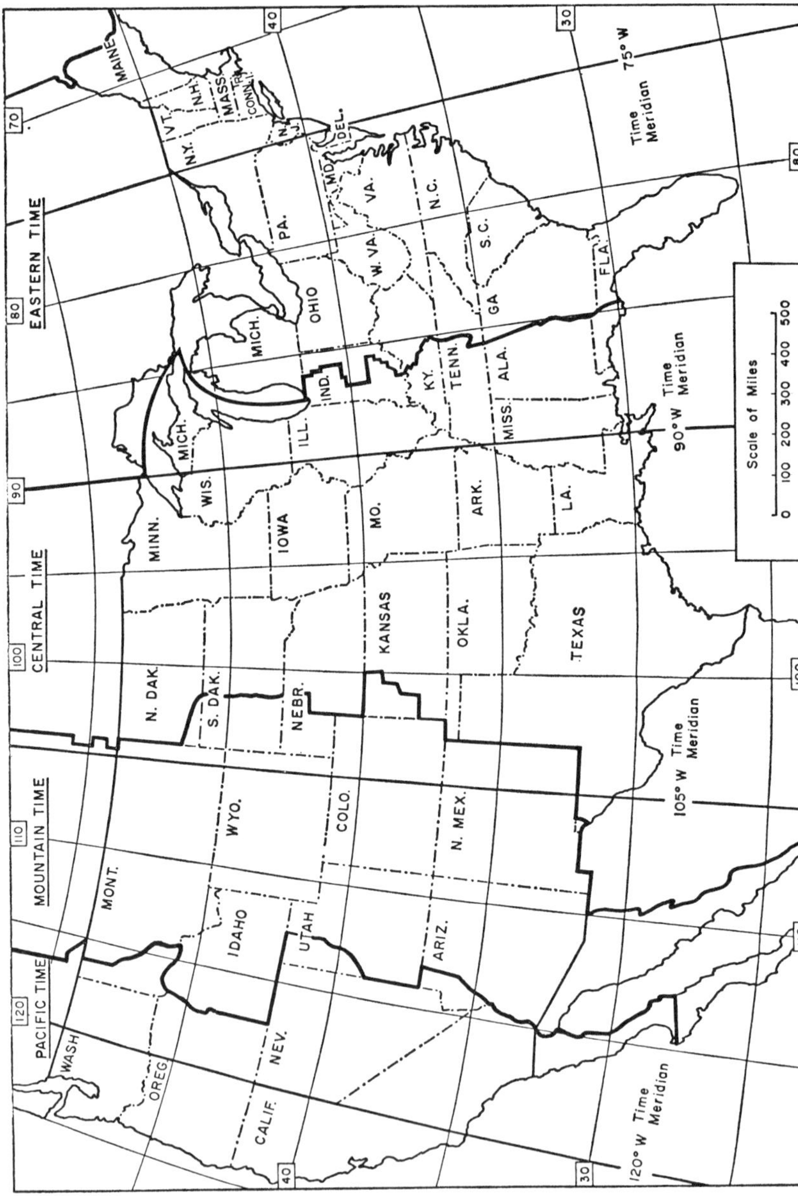
EASTERN TIME
CENTRAL TIME
MOUNTAIN TIME
PACIFIC TIME
75° W Time Meridian
90° W Time Meridian
105° W Time Meridian
120° W Time Meridian
Scale of Miles
0 100 200 300 400 500
MAINE
VT.
N.H.
MASS.
R.I.
CONN.
N.Y.
N.J.
DEL.
MD.
PA.
VA.
W. VA.
N.C.
S.C.
GA.
FLA.
OHIO
MICH.
IND.
ILL.
KY.
TENN.
ALA.
MISS.
WIS.
MINN.
IOWA
MO.
ARK.
LA.
KANSAS
OKLA.
TEXAS
N. DAK.
S. DAK.
NEBR.
WYO.
COLO.
N. MEX.
MONT.
IDAHO
UTAH
ARIZ.
WASH.
OREG.
NEV.
CALIF.
70
80
90
100
110
120
40
30

plain here a few illustrative problems of this nature, and in the following article a number of similar problems are given for readers to solve for their own entertainment and instruction.

Problem 1. Atlanta, Georgia, and Cincinnati, Ohio, are very nearly on the same meridian, and their latitudes are: Atlanta, 33° 45′ N.; Cincinnati, 39° 8′ N. What is the distance from one of the two cities to the other?

Solution. The difference in latitude of the two cities (both North) is $39° 8' - 33° 45' = 5° 23' = 5.38°$. As the length of each degree on the meridian is very nearly 69.2 miles, the required distance is $5.38 \times 69.2 = 372$ miles.

Problem 2. Capetown, South Africa, and Budapest, Hungary, are very near the same meridian, and their latitudes are: Capetown, 34° 21′ S.; Budapest, 47° 30′ N. How far apart are the two cities?

Solution. Since one of the latitudes is North and the other South, the latitude difference, according to article 97, is found by adding the two: $47° 30' + 34° 21' = 81° 51' = 81.85°$. The distance required is therefore $81.85 \times 69.2 = 5{,}664$ miles.

Problem 3. Hollywood, California, and Osaka, Japan, are each within a few minutes of the same parallel of latitude, 34° 30′ N. The longitude of Hollywood is about 118° 45′ W. and that of Osaka is 135° 29′ E. If a ship sails from one of these cities to the other along their parallel of latitude, how far does it travel?

Solution. Along the 34° 30′ parallel the length of one degree of longitude is, by the table in article 95 (proportionally between 30° and 40°; $4\frac{1}{2}°$ in a difference of 10° and 6.7 mi.), equivalent to about 56 miles. Also, as in article 97, the longitude difference between the two cities is $360° - (135° 29' + 118° 45') = 360° - 254° 14' = 105° 46' = 105.8°$, to the nearest tenth of a degree. The required distance (along the parallel) is therefore $105.8 \times 56 = 5{,}925$ miles. (This is not the shortest distance from one of the cities to the other; the shortest distance is along a "great circle" of the earth, article 120.)

Problem 4. An airplane driven by a jet propulsion engine is capable of flying 700 miles per hour. If such a plane could fly 24 hours at that rate without refueling it would travel 16,800 miles. In order to fly around the world along a parallel of latitude in that time, what parallel would it have to follow?

Solution. The parallel (a circle of 360°) must have a length per degree of $16{,}800 \div 360 = 46.7$ miles per degree. In the table of article 95 this parallel is seen to lie between 40° and 50° (N. or S.), and for a difference of 46.7 − 44.4 = 2.3 mi. in a total difference of 52.3 − 44.4 = 7.9 mi. the proportional latitude difference x is found by the proportion $x : 10 :: 2.3 : 7.9$ to be $x = 2.8°$, very nearly. The latitude parallel is therefore 50° − 2.8° = 47.2° = 47° 12′. (In North latitude this parallel passes near the cities of Quebec, Canada; Bismarck, N. Dakota; Seattle, Washington; Rashuwa, Japanese Kuriles Islands; Harbin, Manchuria; Odessa, Russia; Budapest, Hungary; Paris, France; St. Johns, Newfoundland.)

Note. If the plane in the above problem should start at noon and fly westward, the pilot would see the sun stand still in the heavens and he would make the entire trip in daylight, passing near each of the cities named above at its noon time, and return to his starting point with the sun still shining without having set, though his watch would show that 24 hours had passed. In other words, the plane would stand still in space while the earth rotates once beneath it, carrying the atmosphere rushing past the plane at 700 miles per hour. (In order to perform the same feat over the equator the plane must have a speed of 24,912 ÷ 24 = 1038 miles per hour. Starting at noon and flying westward in this case, the pilot would see the sun stand still directly overhead for the entire 24 hours.)

Problem 5. The longitude of Washington, D. C., is 77° 4′ W. and that of Tokyo, Japan, is 139° 45′ E. What is the time difference between the two cities, and what is the local sun (clock) time in Tokyo when the 9.00 A.M. (local sun time) news broadcast begins in Washington?

Solution. The longitude difference between the two cities (one E. and the other W.) is 360° − (139° 45′ + 77° 4′) = 360° − 216° 49′ = 143° 11′. Converting this longitude difference to time difference by Rule I, article 97, (143° 11′ 0″) ÷ 15 = 9 hr. 32 min. 44 sec., or, to the nearest minute, 9 hr. 33 min. The time difference being 9 hr. 33 min. and Tokyo time being ahead of that of Washington, the sun time at Tokyo is 11:27 P.M. of the same day, or 33 minutes before the following midnight.

Problem 6. The longitude of New York City is 73° 58′ 41″ W., or, to the nearest minute, 73° 59′ W., and that of Berlin, Germany, is 13° 22′ E. What is the time difference between the two cities?

Solution. The longitude difference is 73° 59′ + 13° 22′ = 87° 21′

(less than 180°), and the time difference is therefore (87° 21′) ÷ 15 = 5 hr. 49.4 min., or to the nearest minute, 5 hr. 49 min., Berlin time being ahead of New York City time.

101. Problems for Solution.

1. The southernmost tip of Texas is in latitude about 26 N. and the Canadian boundary line is latitude 49° N. How wide is the United States?

2. New York City is in about 41° N. latitude and Buenos Aires about 35° S. What is the north-south distance between these cities?

3. The easternmost tip of Maine is at 45° N., 67° W. and the Pacific Coast of the United States in that latitude is at longitude 124° W. What is the east-west width of the United States in that latitude?

4. The Suez Canal and the city of New Orleans are both on the 30° N. parallel of latitude, New Orleans being at longitude 89° W. and Suez $31\frac{1}{4}$° E. How far apart are they in miles along their parallel?

5. When it is 9 o'clock at Washington (true time) it is 8:07:04 at St. Louis, and the longitude of Washington is 77° 1′ W. What is the longitude of St. Louis?

6. The sun rises at Boston 1 hr. 11 min. 56 sec. earlier than at New Orleans, the longitude of New Orleans being exactly 89° 2′ W. What is the longitude of Boston?

7. When it is 2:30:00 A.M. at Havana, it is 9:13:20 A.M. at the Cape of Good Hope, the longitude of the latter being 18° 28′ E. What is the longitude of Havana?

8. A ship's clock, set at Greenwich, shows the time to be 5 hr. 40 min. 20 sec. P.M. when the sun is on the meridian at the same place. What is the longitude of the ship's position?

9. Washington is 77° 1′ W. and Cincinnati 84° 24′ W. What is the true time difference between the two places?

10. Buffalo is 78° 55′ W. and Rome, Italy, 20° 30′ E. What is their time difference?

11. The longitude of Cambridge, Mass., is 71° 7′ W. and that of Cambridge, England, is 5′ 2″ E. When it is noon at Cambridge, England, what time is it at Cambridge, Mass.?

12. The longitude of New York City is, to the nearest minute, 73° 59′ W., and that of Manila, Philippine Islands, is 120° 59′ E. What is the (sun) time difference between the cities?

13. Washington, D. C., and Lisbon, Portugal, are both very near the 38° 49′ N. parallel of latitude, and their longitudes are: Washington, 77° 4′ W.; Lisbon, 9° 11′ W. What is the length in miles of the latitude parallel joining the two cities?

14. The longitude of Leningrad, Russia, is 30° 18′ E., and the longitude of the eastern edge of the Kamchatka Peninsula of Russia at the same latitude (60° N.) is 165° E. What is the distance across Russia along that parallel of latitude?

15. Algiers, Algeria, and Paris, France, both lie within a few minutes of the same meridian of longitude, and their latitudes are: Algiers, 36° 48′ N.; Paris, 48° 50′ N. How far apart are the two cities?

16. Boston, Mass., and Rome, Italy, are both very near the 42° 8′ N. parallel of latitude, and their longitudes are: Boston, 71° 4′ W.; Rome, 12° 29′ E. Find the time difference between the two cities, and the distance between them along their parallel.

17. The American aviators Wiley and Post in their "round the world" trip did not fly around the greatest circumference of the earth but approximately along or near the 55° N. parallel. What distance did they cover, approximately?

Chapter 16

DIMENSIONS AND AREAS OF PLANE FIGURES

102. Introduction. In this chapter we take up the study of the so-called *geometrical figures* such as squares, triangles, circles, spheres (globes), etc., which are much used in drawing, building, the trades, arts and sciences. We shall assume that the reader has some knowledge of the names and appearances of the most common of these figures and shall not attempt to give precise theoretical mathematical definitions of them, but shall give simple descriptive definitions in ordinary language.

These figures and their properties are much used in such work and subjects of study as drawing, painting, plastering, papering, building operations, land measure, capacity of bins, tanks, measuring vessels, and the like. For this reason such subjects and calculations are often treated in books on arithmetic. They more properly belong in books on the trades or on vocational arithmetic, however, and will not be treated here. We shall, therefore, treat simply of the arithmetical *properties* of such figures, their forms, dimensions, areas and volumes, and leave to other books the *uses* of these properties.

Geometrical figures are of two kinds, called flat or *plane*, and *solid;* and among the solid figures there are two kinds again, those which have plane surfaces and those which have curved or rounded surfaces. This chapter will only treat of plane figures; solid figures will be treated in the next chapter.

103. Kinds and Properties of Plane Figures. A *plane* figure is one which has or is drawn upon a flat surface such as the top of a smooth table, and is made up of straight or curved lines and angles. Such figures are the familiar square, triangle and circle, as well as others perhaps not so familiar.

There are many different forms of plane figures and all have characteristics peculiar to themselves. These can be classified in groups, however, and studied in general from the same viewpoint. Thus, plane figures formed of straight lines are generally named and studied with reference to the number of straight sides or the number of angles they possess. Of the great number thus formed a few are of greater importance or usefulness than the others.

Thus, the three-sided or three-angled figures are called *triangles*, four-sided figures are *quadrilaterals*, five-sided figures are *pentagons*, six-sided figures are *hexagons*, and so on. In general, figures of several or many sides are called *polygons*, and if a polygon has all its angles equal and all its sides of equal length it is called a *regular* polygon.

In particular, if all four of the angles of a quadrilateral are right angles it is called a *rectangle*, and if the four sides of a rectangle are equal it is called a *square*.

If the opposite sides in pairs of any quadrilateral are parallel the figure is called a *parallelogram*. If only two of the sides of a quadrilateral are parallel it is called a *trapezoid*.

An important *round* plane figure is the *circle*. Its chief and distinguishing feature is this roundness and circles can differ only in size, but the circle has many very interesting properties and for thousands of years has been the object of interested study.

The lengths of the boundary lines or "edges" of plane figures, and of their other characteristic or distinguishing lines, are called their *dimensions*. Dimensions are measured in length or *linear* measure, as given in the tables of Chapter 12.

The extent or amount of surface enclosed within the boundary lines of a closed plane figure is called its *area*. Area is measured in *square* measure, as given in Chapter 12.

There are many interesting relations between the dimensions and areas of plane figures, generally such that when some are known for a particular figure the others can be calculated. These relations and various other characteristics of plane figures are called their *properties*.

The complete study of the properties of plane figures forms a separate branch of mathematics, the subject called *plane geometry*. In arithmetic we study the most useful properties of the most commonly used figures and the methods of calculating their areas and dimensions.

104. The Rectangle and Square. In Fig. 14 the figure with the angles or corners marked *ABCD* is a *rectangle*, having four straight sides and four right angles. The side *AB* which equals *CD* is called the *length* or the *base*, and the side *BC* which equals *AD* is the *width*, or *altitude* if the rectangle is supposed to stand on its base.

In Fig. 14(b) suppose the length or base $AB = 5$ inches and the width or altitude $BC = 3$ inches. Each of the small squares into which the rectangle is divided is then 1 inch on a side and is called a *square inch.* The small shaded square then represents 1 sq. in. In

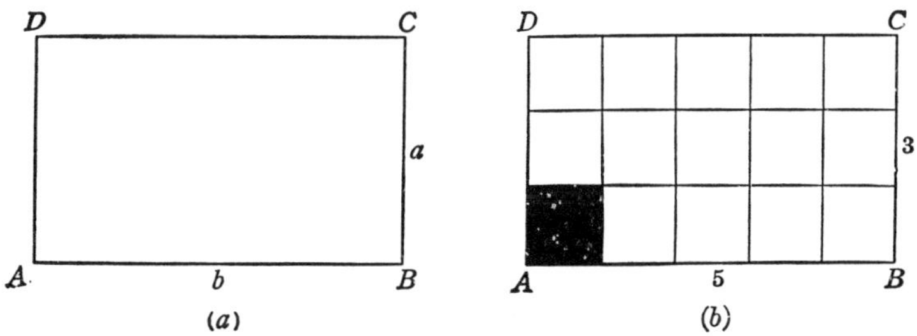

FIG. 14.

the whole surface or *area* of the rectangle there are 3 rows of 5 square inches each, or 5 rows of 3 square inches each. The area of the rectangle is therefore 3×5 or $5 \times 3 = 15$ square inches. That is, the area is the product of the number of units in the length and the number of units in the width, or the number of units in the base and number of units in the altitude. The same relation holds true for any rectangle. We have, therefore, the result that

The area of a rectangle is the product of the number of units in its length and the number of units in its width.

Since, in division, the dividend is the product of the quotient and divisor, we have, therefore, from the area rule that

Either dimension of a rectangle is the quotient of the area by the other dimension.

If, as in Fig. 14(a), we let a represent the number of units in the width or altitude and b the number of units in the length or base, and also let A represent the area, then

$$A = a \times b.$$

A square is a rectangle with all four sides equal, that is, the length is the same as the width. The product of the two is, therefore, the square of the side. This gives the rule that

The area of a square is the square of the length of its side.

Reversing this rule, we have that

The side of a square is the square root of its area.

Using the same symbols as in the case of the rectangle, these rules can be written $A = a \times a$ or

$$A = a^2, \text{ and } a = \sqrt{A}.$$

Thus, if a square is 5 inches on a side the area is $A = 5^2 = 25$ sq. in. If the area of a certain square is 36 sq. ft., then the side is $a = \sqrt{36} = 6$ ft.

105. The Parallelogram. The figure $ABCD$ in Fig. 15(a) is a parallelogram, having the two opposite sides AB and CD parallel and also equal in length, and similarly the sides AD and BC. The line AB which is equal to CD is called the *base* of the parallelogram and the perpendicular DE, and NOT the length of AD or BC, is called the *altitude*. The length DE is simply the direct distance between the base and its opposite parallel side.

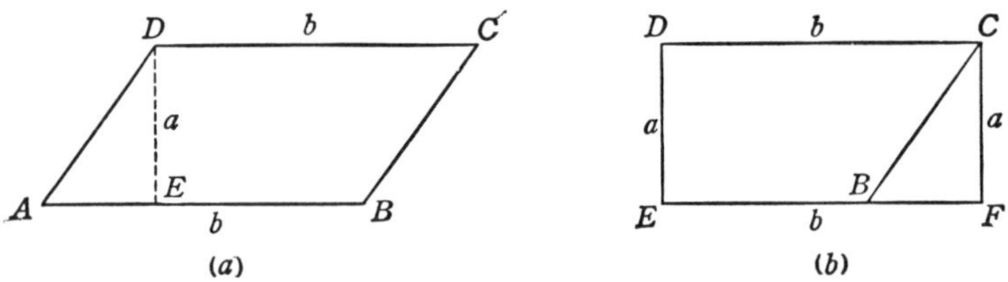

FIG. 15.

If the section DAE in Fig. 15(a) is removed and placed at the other end of the parallelogram, as at CBF in Fig. 15(b), then the parallelogram becomes a rectangle with the same base and altitude and, as in the preceding article, the area is the product of the base and altitude. Therefore,

The area of a parallelogram is the product of the number of units in its base and the number of units in its altitude.

If, in Fig. 15(a), we let the base $AB = b$ and the altitude $DE = a$, then, as before, the area is

$$A = a \times b.$$

Care must be taken not to confuse the altitude with the slant width of the parallelogram.

106. Triangles. A *triangle* is a plane figure with three straight sides and three angles. The figure *ABC* in Fig. 16(*a*) is a triangle. Triangles may have any shape whatever, as in Fig. 16(*a*), (*b*), (*c*) and (*d*).

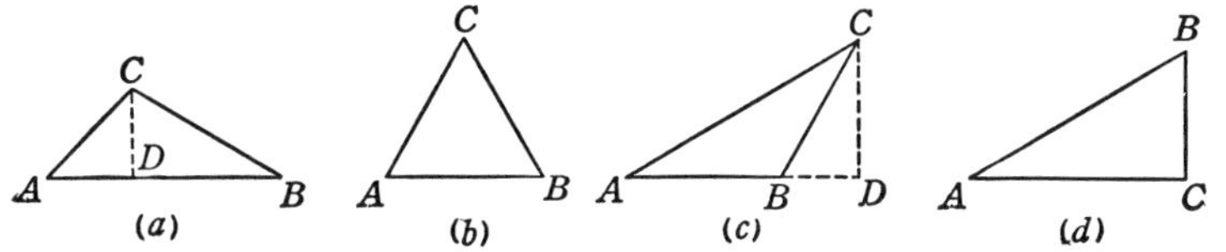

FIG. 16.

If the three sides are of equal length, the triangle is called *equilateral* or *equiangular*, since the angles are in that case also equal. Fig. 16(*b*) represents an equilateral triangle. If there is no regularity among the sides and angles of a triangle, and one of the angles is an obtuse angle (greater than a right angle), the triangle is called an obtuse or *oblique* triangle; *ABC*, Fig. 16(*c*) is such a triangle, as is Fig. 16(*a*). If one of the angles of a triangle is a right angle, as at *C* in Fig. 16(*d*), the triangle is called a *right triangle.*

Each of the three "corners" or angle "points" of a triangle is called a *vertex* of the triangle. The side of a triangle upon which it is represented as standing is called its base. Thus, in Fig. 16(*a*), (*b*) and (*c*) the side *AB* is the base and in (*d*) the base is *AC*. A line drawn perpendicular to the base from the opposite vertex, as *CD* in Fig. 16(*a*) and (*c*), is called the *altitude* of the triangle. In the right triangle, Fig. 16(*d*), since the side *BC* is already perpendicular to the base *BC*, it is itself the altitude.

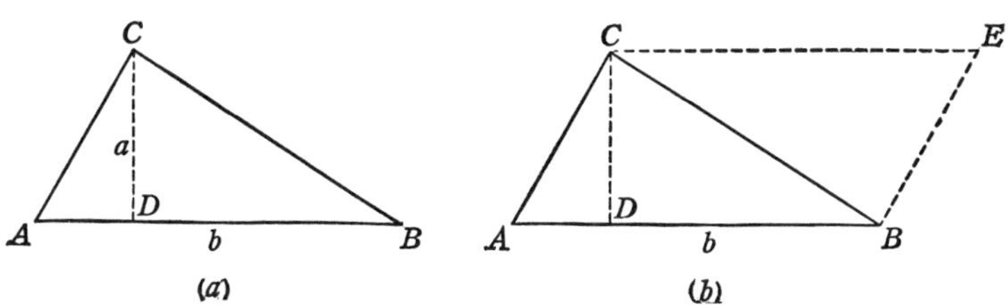

FIG. 17.

Any triangle, such as *ABC* in Fig. 16(*a*) or 17(*a*), may be thought of as forming half of a parallelogram with the same base and altitude, as in Fig. 17(*b*). Then, since the area of the parallelogram is the product of the number of units of the base and number of units of the altitude, that of the triangle will be *half* this product. That is,

The area of any triangle is half the product of number of units in its base and number of units in its altitude.

This rule holds good for triangles of any shape and if, as in Fig. 17(*a*), we again let *a* represent altitude and *b* base, we can write for all triangles that the area is

$$A = \tfrac{1}{2}(a \times b).$$

From this result, triangles of different shapes may have the same area, so long as they have the same base and altitude or the product of the base and altitude is the same.

Another very useful rule for finding the area of a triangle when the three sides are known instead of the base and altitude is the following:

From half the sum of the sides subtract each side separately; multiply the half-sum and the three remainders together; the square root of the product is the area.

As an example of the two rules for the area of a triangle, consider a triangle such as *ABC*, Fig. 17(*a*), in which the sides *AB*, *BC*, *CA* are, respectively, 25, 20, 15 inches. The altitude is then found to be $CD = 12$ inches. By the first rule, the area is

$$A = \tfrac{1}{2}(12 \times 25) = 150 \text{ sq. in.}$$

By the second rule, half the sum of the sides is $\frac{1}{2}(25 + 20 + 15) = 30$, and the remainders after subtracting each side in turn from this half-sum are $30 - 25 = 5$, $30 - 20 = 10$ and $30 - 15 = 15$. According to the rule, the area is then

$$A = \sqrt{30 \times 5 \times 10 \times 15} = \sqrt{22{,}500} = 150 \text{ sq. in.},$$

which is the same as that obtained by using the first rule.

The choice of rule to use will depend on which dimensions of the triangle are known in any particular case.

Another fundamental and in some kinds of work very useful property of triangles is that

The sum of the three angles of any triangle is 180 *degrees.*

This is proved in plane geometry and is much used in the study of the general properties and measurement of triangles, called *trigonometry*. In this general study of triangles, the right triangle is of supreme importance, as is also the case in many applications in building, surveying, measurements, etc.

107. The Right Triangle. As we have seen, a *right triangle* is a triangle which has one right angle, that is, one angle of 90°. Then since the sum of all three angles is 180°, the sum of the other two is 90°. Each of these two must, therefore, be less than 90° and so is an *acute* angle.

The sides of a right triangle which form the right angle are called its *legs* and the third or diagonal side, opposite the right angle, is called the *hypotenuse*. When a right triangle stands on one of its sides as a base, then the other is the altitude, as in Fig. 16(*d*). The area rule when applied to a right triangle becomes, therefore:

The area of a right triangle is half the product of the number of units in its legs.

The lengths of the two legs of a right triangle, altitude and base, are denoted by a and b and the hypotenuse by c. Thus, in Fig. 18 the legs and hypotenuse of the right triangle with right angle at C are respectively a, b, c. It is to be noted that the side c is opposite the right angle at C and the sides a and b are opposite the corresponding acute angles A and B. This is the standard method of drawing and lettering right triangles.

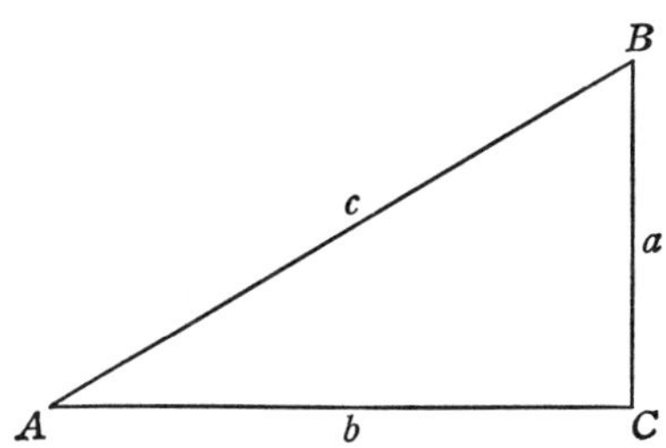

FIG. 18.

A very important property of the right triangle is that

The square on the hypotenuse equals the sum of the squares on the legs.

This relation is illustrated in Fig. 19 where the legs a, b of the right triangle ABC are, respectively, 3 and 4 units and the hypotenuse c is 5 units long (these may be inches, centimeters, feet, or any other linear unit). The areas of the squares on the legs are then $3^2 = 9$ and $4^2 = 16$ square units, and the area of the square on the hypotenuse is $5^2 = 25 = 9 + 16$.

The truth of the relation just stated can be seen from an examination of Fig. 20. Here the squares on the legs of the right triangle *ABC* are the squares *BCGH* and *ACED*, while the square with the hypotenuse *AB* as side is the square drawn in dotted lines. The large square *DFHI* is made up of the squares on the legs and the two equal rectangles *CEFG* and *ACBI*. But the rectangle *ACBI* is equal to the two triangles *ABC* and *ABI* which are the same. Therefore, the *four* equal triangles 1, 2, 3, 4 are equivalent to the *two* rectangles.

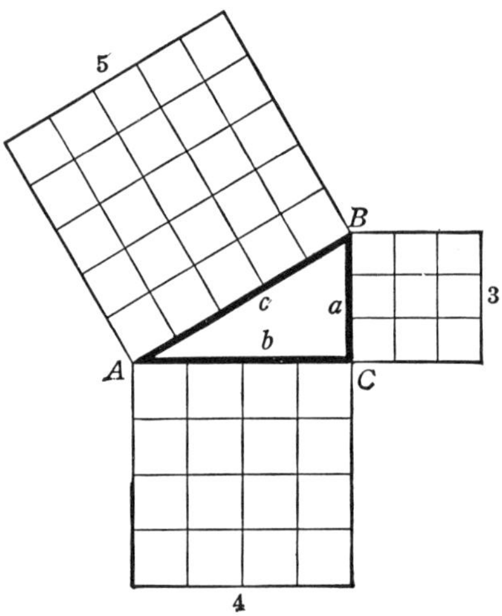

FIG. 19.

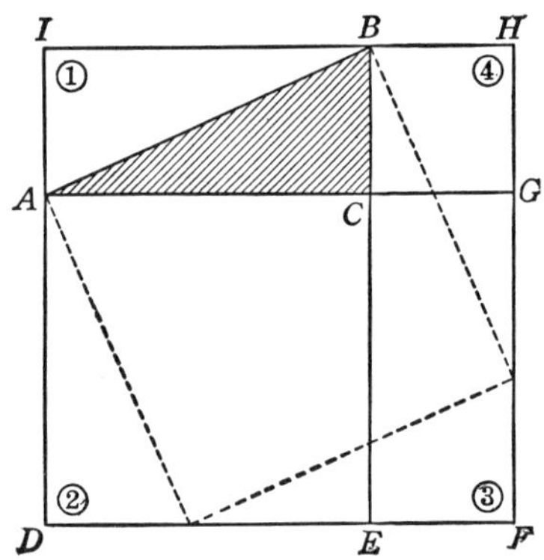

FIG. 20.

If now the four triangles are removed from the large square *DFHI*, there remains the dotted line square on the hypotenuse, and if the two rectangles are removed from the large square, there remain the two squares on the legs *AC* and *BC*. But, as already seen, the four triangles are equal in area to the two rectangles. Therefore, the two remainders are equal, that is, the square on the hypotenuse is equal to the sum of the squares on the legs, as was stated.

Now the areas of the squares with the legs a, b as sides are a^2 and b^2 and the area of the square with the hypotenuse c as side is c^2. Therefore, since the area of the square on the hypotenuse is the sum of the other two, we can write in symbols

$$c^2 = a^2 + b^2,$$

which, in words, referring to the *lengths* of the sides, is

The square of the hypotenuse of any right triangle equals the sum of the squares of the legs.

From this we have at once that

The square of either leg of a right triangle equals the difference of the squares of the hypotenuse and other leg.

These two results can be expressed as the following rules:

RULE I. *To find the hypotenuse of a right triangle when the legs are known: Square the lengths of the legs and add the results. The square root of the sum is the hypotenuse.*

RULE II. *To find a leg of a right triangle when the other leg and the hypotenuse are known: Square the lengths of the hypotenuse and the known leg and subtract the results. The square root of the remainder is the required leg.*

As an example of these rules, consider the triangle *ABC* of Fig. 19. We have already seen that the two legs when squared and added give $9 + 16 = 25$. The hypotenuse is then by Rule I equal to $\sqrt{25} = 5$. If this hypotenuse and either leg, say the 3, were known, but the other leg unknown, then, by Rule II, we would find $5^2 = 25$ and $3^2 = 9$; subtract, $25 - 9 = 16$; and the other leg is then $\sqrt{16} = 4$.

As another example, let us find the hypotenuse of the right triangle whose legs are 8 inches and 6 inches. By Rule I: Squaring, $8^2 = 64$, $6^2 = 36$; adding, $64 + 36 = 100$; taking square root, $\sqrt{100} = 10$ inches, the hypotenuse.

Suppose the hypotenuse of a certain right triangle is 13 yds. and one leg is $10\frac{2}{5}$ yds. long; how long is the other leg? First, express the fraction as a decimal, $10\frac{2}{5} = 10.4$. Then, by Rule II: Squaring, $13^2 = 169$, $10.4^2 = 108.16$; subtracting, $169 - 108.16 = 60.84$; taking square root (by logarithms as in article 39), $\sqrt{60.84} = 7.8 = 7\frac{4}{5}$ yd., the other leg.

The relation between the legs and the hypotenuse of certain right triangles has been known since very ancient times, but it was first definitely proved to be true for *any* right triangle by a Greek mathematician and philosopher named Pythagoras, about 500 B.C. It is called after him the rule or *Theorem of Pythagoras* and is a very famous rule in the history of mathematics. The proof or explanation which is given above in connection with Fig. 20 is thought to have been the one given by Pythagoras himself.

As indicated before, the properties of right triangles play a large and important part in that branch of mathematics called *trigonometry*, which has to do with the general study and measurement of triangles, or *trigons*.

108. The Trapezoid. A plane figure of four sides, two and only two of which are parallel, is called a *trapezoid*. The figure *ABCD*, Fig. 21(*a*), is a trapezoid, the two sides *AB* and *CD* being parallel. The two parallel sides of a trapezoid are called its two *bases* and the direct or perpendicular distance between these two is called its *altitude*. In Fig. 21(*a*) the bases are *AB* and *CD* and the altitude is *GH*.

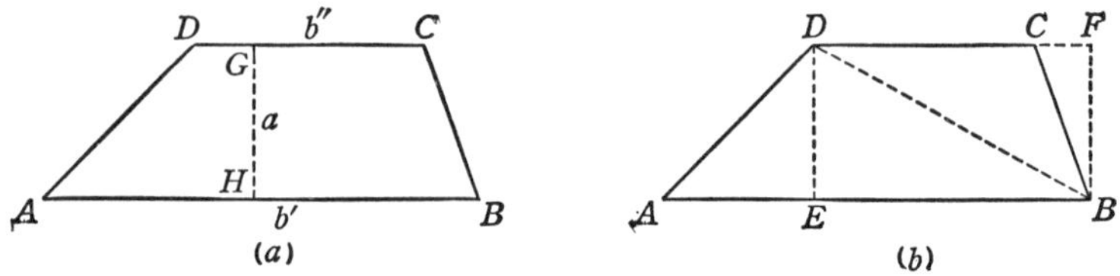

FIG. 21.

As shown in Fig. 21(*b*), the trapezoid *ABCD* can be considered as formed of two triangles, *ABD* and *BCD*, whose bases *AB* and *CD* are the bases of the trapezoid and whose altitudes *DE* and *BF* are both the same as the altitude of the trapezoid. The area of each of these triangles is half the product of its base and altitude and the area of the trapezoid is the sum of these two areas. The area of the trapezoid is, therefore, half the product of its altitude and one base plus half the product of the altitude and the other base. That is,

The area of a trapezoid is half the product of its altitude by the sum of its bases.

If we let a represent the altitude, b' one of the bases, and b'' the other ("*b* prime" and "*b* second," not minutes and seconds) then the area can be written in symbols

$$A = \tfrac{1}{2}a \times (b' + b'').$$

Thus, in Fig. 21(*a*), if the base *AB* is $b' = 12$ inches, base *CD* is $b'' = 8$ inches, and altitude *GH* is $a = 4$ inches, then the area is

$$A = \tfrac{1}{2}\,4 \times (12 + 8) = \tfrac{1}{2}\,4 \times 20 = 40 \text{ sq. in.}$$

109. The Circle. The familiar circle is in form the simplest of the plane figures but in interest and importance it ranks with the triangle, having been the object of study for thousands of years and forming the basis of many subjects in mathematics as well as the basis of many practical constructions and applications.

Fig. 22 represents a *circle*. The smoothly curving line forming the figure of the circle is called its *circumference*, and every point on the circumference is at the same distance from the point O inside, which is called the *center*. The length of the circumference is denoted by the initial letter C. Any straight line passing through the center and having its two ends on the circumference, such as the line AB, is called a *diameter* of the circle, and any line from the center to any point on the circumference such as the line OP, is called a *radius*. Since OA and OB are, therefore, radii (plural), the length of the diameter is twice the length of the radius, or the radius is half the diameter. Denoting the radius by R and the diameter by D, therefore, we have that $D = 2R$ and $R = \frac{1}{2}D$.

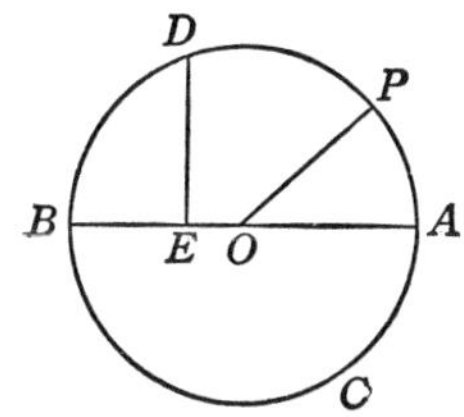

FIG. 22.

If both the circumference and the diameter of each of any number of circles are very carefully measured and the length of the circumference of each is divided by the length of its own diameter, it will be found that the quotient is in every case, regardless of the size of the particular circle, very nearly $3\frac{1}{7}$, and if a large number of such quotients be found and the average be calculated (by adding them and dividing the sum by the number of them) it will be found to be very nearly 3.1416, which is more nearly exact than $3\frac{1}{7}$.

In plane geometry it is proved that this result is correct for *any* circle, of any size. It is further proved, however, that this number is a never-ending decimal, but no matter how far it may be carried, that is, to no matter how many decimals, it is always the same for every circle. To 10 decimal places it is 3.1415926535 and it has been calculated by methods of higher mathematics to over 700 decimal places. This so-called "transcendental" number is, for convenience in writing, usually represented by a single symbol, the Greek letter π ("pi," pronounced as "P" or, more commonly, mispronounced as "pie").

Much of the interest in the circle has centered around the value of π and for thousands of years mathematicians and others of all races attempted to find its value. If it could be represented by a definite decimal with a specified number of decimal places, or by a common fraction with a definite numerator and denominator, it would be possible to find the dimensions of a square, that is, the side of the square, with exactly the same area as that of a circle of any specified size, or diameter. Thc problem of finding the value of π is, therefore, referred to as the problem of "squaring the circle." It was not until the year 1768 that π was definitely proven to have no such exact value, and therefore, that it is impossible to "square the circle."

Since it is true that for any circle, the length of the circumference divided by that of the diameter gives the quotient $\pi = 3.1416$ to a very close approximation, we can say that for all practical purposes,

The circumference of any circle is 3.1416 *times the diameter; and the diameter is the circumference divided by* 3.1416.

This rule is very useful in all calculations which involve the circle. Using the symbols for circumference and diameter given in connection with Fig. 22, and the symbol $\pi = 3.1416$, we can write

$$C = \pi \times D, \text{ and } D = C \div \pi.$$

Thus, if the diameter of a certain circle is 10 inches, the circumference is $3.1416 \times 10 = 31.416$ inches or nearly $31\frac{1}{2}$ inches. If the diameter is 1 foot, the circumference is $3.1416 \times 1 = 3.1416$ or nearly $3\frac{1}{7}$ feet, a little over 1 yard. (In the early days when it was thought that $\pi = 3$, this may have been the origin of the yard as 3 feet; the fence around a circular "yard" being three times the distance across the center.)

Since the diameter of the circle is twice the radius,

The circumference of a circle is 6.2832 *times the radius; and the radius is the circumference divided by* 6.2832.

Using symbols as before, this can be written

$$C = 2\pi \times R, \quad R = C \div 2\pi,$$

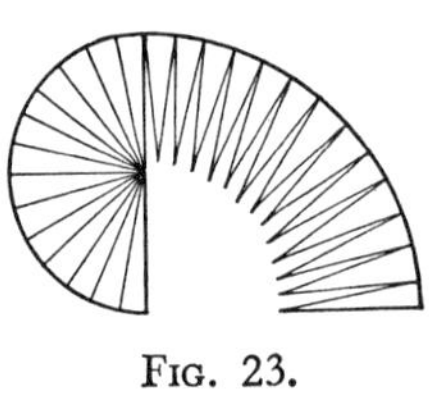

FIG. 23.

where 2π is $2 \times 3.1416 = 6.2832$.

The circle may be divided into a large number of narrow "pieces of pie" which are essentially triangles, as in Fig. 23. The altitude of each of these triangles is the radius of the circle, the sum of their bases is the circumference of the circle, and the sum of their areas is the area of the circle. The area of each being half the product of the base and altitude, the sum of the areas is the sum of the separate products, or half the product of the *altitude* (same for all) by the *sum of the bases.* That is,

The area of a circle is half the product of its radius and circumference.

Since the circumference is 2π times the radius, this rule gives: *area* $= \frac{1}{2}[2\pi \times (radius) \times (radius)]$, that is $\pi \times (radius)^2$. In words this is:

The area of a circle is 3.1416 *times the square of its radius.*

Since the radius is half the diameter, the square of the radius is one fourth the square of the diameter. Therefore, the area is $\frac{1}{4} \times 3.1416 \times (diameter)^2$, or, since $\frac{1}{4} \times 3.1416 = .7854$,

The area of a circle is .7854 *times the square of its diameter.*

These two rules allow the area of a circle to be calculated if either the radius or the diameter alone be known. In symbols they can be written

$$A = \pi \times R^2, \quad \text{or} \quad A = \frac{\pi}{4} \times D^2.$$

Thus, the area of a circle of 5 inch radius, or 10 inch diameter, is $3.1416 \times 5^2 = 3.1416 \times 25 = 78.54$ sq. in., or $.7854 \times 10^2 = .7854 \times 100 = 78.54$ sq. in., the same.

If the area of a circle is known, the diameter and radius are easily found by reversing the last two rules, as follows:

RULE: *To find the diameter or radius of a circle whose area is known: Divide the area by* .7854, *and the square root of the quotient is the diameter, and the radius is half the diameter;* or *divide the area by* 3.1416, *and the square root of the quotient is the radius, and the diameter is twice the radius.*

Among the great number of interesting geometrical properties which the circle possesses, we shall mention one more which is of interest in arithmetic. If, in the circle of Fig. 22, a perpendicular such as the line *DE* be drawn from any point *D* in the circumference to the diameter *AB*, then the length of *DE* is a mean proportional of the lengths *BE* and *EA* into which *DE* divides the diameter. As in article 62, therefore, the length of *DE* is the square root of the product of the lengths *BE* and *EA*. Thus, if the diameter of a circle is 15 inches and the perpendicular *DE* is drawn in such a position that *BE* is

3 inches and *EA* 12 inches, then by the rule for the mean proportional, the length $DE = \sqrt{3 \times 12} = \sqrt{36} = 6$ inches.

This relation holds true for any position of *DE.* It is because of this proportional relation (which is proved in geometry) that the mean proportional is called the *geometric mean.*

110. Problems.

1. The side of a square park is 495 yards. What is its area in acres?
2. The area of a square is 17,424 sq. in. How long is its side in feet?
3. At $2.80 per sq. yd. what is the cost of paving a triangular courtyard, its base being 105 ft. and altitude 21 yards?
4. Find the area of the gable end of a house that is 28 feet wide with the ridge of the roof 15 feet higher than the foot of the rafters.
5. Find the area of a triangle whose sides are 30, 40, 50 feet.
6. How many acres are there in a field of the form of an equilateral triangle whose sides each measure 70 rods?
7. The roof of a house 30 feet wide has the rafters on one side 20 feet long and on the other 18 feet. How many square feet of lumber are required to board up both gable ends?
8. The base of a right triangle is 12 inches and the other leg is 16 inches. How long is the hypotenuse?
9. The foot of a ladder cannot be placed closer than 15 feet from the wall of a building and must reach a window ledge 36 feet above the ground. How long must the ladder be?
10. If the gable end of a house is 16 feet above the attic floor, which is 40 feet wide, the rafters on both sides being of equal length, what is that length?
11. A room is 20 feet long, 16 feet wide, and 12 feet high. What is the length from one of the lower corners to the opposite upper corner?
12. A ladder 52 feet long leans against the side of a building with its foot 20 feet from the wall. How many feet must it be drawn out at the bottom in order that the upper end may be lowered 4 feet?
13. Find the circumference of a circle whose diameter is 20 inches.
14. Find the height of a wheel standing upright if the circumference is 50 feet.
15. What is the diameter of a tree whose girth is 18 ft. 6 in.?
16. Find the length of tire necessary to band a wheel 7 ft. 9 in. in diameter.
17. An automobile wheel stands $2\frac{1}{2}$ feet, top to bottom. How many times will it turn over in going a mile without slipping?
18. Find the area of a circle whose diameter is 10 feet and circumference 31.416 feet.
19. Find directly the area of a circle of diameter 12 inches.
20. The distance around a circular park is $1\frac{1}{2}$ miles. How many acres does it contain?
21. Find the area of the largest circle that can be drawn by using as a radius a string 20 inches long.

22. Find the area of a circular ring when the diameters of the inner and outer bounding circles are 20 and 30 feet.

23. A walk 4 feet wide borders a circular pond which measures 50 feet across. What is the cost of paving the walk at $3.75 a sq. yd.?

24. A circular flower bed is at the center of a square lawn, the remainder of which is covered in grass. If the area of the circle is 95 sq. ft. and the side of the lawn is 31 feet, how far is the edge of the flower bed from the side and from the corner of the lawn and what area of the lawn is in grass?

25. As an airplane flying at a height of 4500 feet passes over a certain landmark it is observed by a person who is $2\frac{1}{2}$ miles from the landmark on the same level with it. How far is the plane from the observer?

Chapter 17

DIMENSIONS, AREAS AND VOLUMES OF SOLIDS

111. Introduction. A *solid figure* is one which encloses or occupies a portion of space, as, for example, a box, a block of stone, a tin can, a globe or ball, and the like. Such a figure may be "solid" or "hollow" in the ordinary sense of these words, and it may be closed or open, filled or empty. In all cases it will be called a *solid* to distinguish it from a *plane* figure, and its properties will be studied as if it were as "solid" as a block of stone.

There are many different forms of solids. They may be classified as those which have plane surfaces and straight edges and those which do not. The surfaces of the first kind of solid may be any of the plane figures studied in the preceding chapter. The surfaces of the second kind of solids may be curved in any way whatever. Altogether the number of forms of solids is unlimited.

Of this great number of possible solids, however, only a few are of common interest or practical importance and only these will be studied here.

112. Kinds and Properties of Solids. Of the solids whose surfaces are made up of plane figures, among the most important are those for which these plane figures are rectangles. These are called *rectangular* solids, and of the rectangular solids an extremely important one is that whose surface is formed of *squares*. Such a solid is called a *cube*. Solids whose surfaces are formed of triangles, with possibly one other figure, are also important and are called *pyramids*.

The different portions of the surface of a solid, each of which is a single plane figure, are called its *faces*. Thus, a cube has six faces, each of which is a square.

If the faces of a solid are regular polygons, all of which are alike and of the same size, it is called a *regular* solid. There are only a certain number of regular solids and no more.

Of those solids whose surfaces are not formed of planes, the most important are the so-called "three round bodies," the *cone*, the *cylinder*, and the familiar *sphere*.

Solids with plane faces and the three round bodies will be treated in this chapter in separate sections.

As in the case of plane figures, the lengths of the edges and curved lines which bound a solid, and of other distinguishing or characteristic lines on or in the figure, are called its *dimensions*. The area of its faces or the sum of those areas, or the area of the curved surface of a solid, is called its *surface area*. The amount of space which it occupies or encloses is called its *volume*.

As in the case of plane figures, the dimensions and area of solids are measured in linear and square measure, respectively. Volume is expressed in *cubic* measure, as given in the tables of Chapter 12.

The characteristic features of solids and the relations between their dimensions, areas and volumes are collectively referred to as their *properties*. The full study of the properties of solids forms a separate branch of mathematics, the subject called *solid geometry*.

A. SOLIDS WITH PLANE FACES

113. Rectangular Solids. A *rectangular solid* is a solid which has six plane faces, each of which is a rectangle. Such an object as a brick, or a block, or a simple box with straight edges and square corners, is a rectangular solid. Fig. 24 represents a rectangular solid.

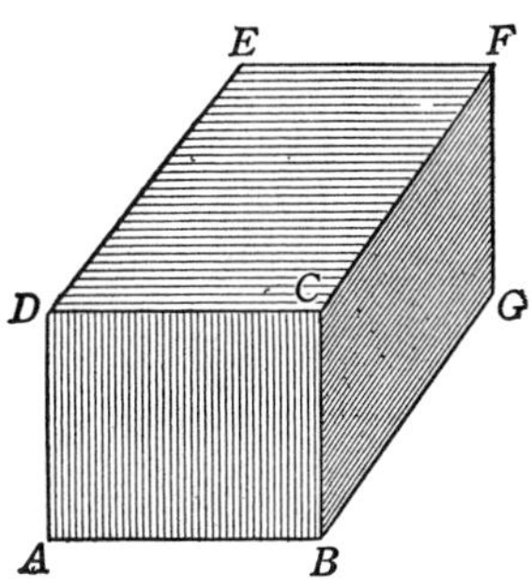

FIG. 24.

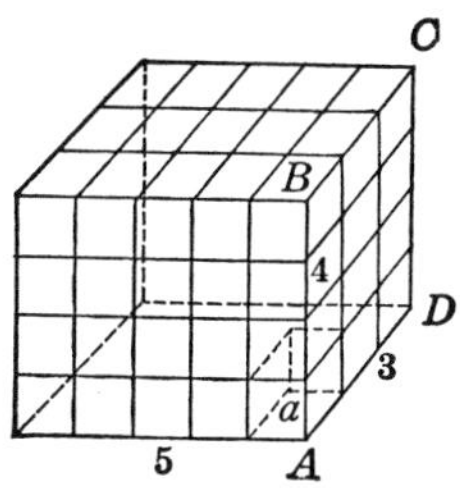

FIG. 25.

Only three of the six faces of a rectangular solid can be seen at any one time. In Fig. 24 these three visible faces are the rectangles *ABCD*, *BCFG*, and *CDEF*. These three faces may be called the end, side and top, respectively.

A rectangular solid has three dimensions: its *length*, *width* (or breadth) and *depth* (height or thickness). In Fig. 24 $BG = CF = DE$ is the length, and may be represented by l; $AB = DC = EF$ is the width, represented by w; and $AD = BC = GF$ is the depth d. The lines named are also called the *edges*, of which the figure has twelve, at least three being invisible in any view.

Of the six faces there are three equal pairs: top and bottom, two sides, and two ends. The area of the top is the product of length and width ($l \times w$), and of the top and bottom, twice this product. The area of either side is the product of length and depth ($l \times d$), and of both sides, twice this. Similarly, the area of the two ends is twice the product of width and depth ($w \times d$). The total surface area is the sum of these double products or double the sum of the three products. That is,

The total surface of a rectangular solid is twice the sum of the three products: length by width, width by depth, depth by length.

In symbols this is

$$A = 2[(l \times w) + (w \times d) + (d \times l)].$$

Suppose that the dimensions (lengths of edges) of a rectangular solid are 3, 4, and 5 inches, as shown in Fig. 25, and imagine the solid built of square blocks 1 inch on each edge such as a shown at the corner A. This little square block is 1 *cubic inch.* The section $ABCD$ is then a layer consisting of $3 \times 4 = 12$ of these blocks, or cubic inches, and the whole space occupied by the solid, its *volume*, is filled up by 5 such layers as $ABCD$. There are, therefore, $5 \times 12 = 60$ of these blocks, that is, 60 cu. in., in the entire solid. That is, it contains, or its volume equals, 60 cu. in. But $60 = 5 \times 12 = 5 \times 4 \times 3$ is the product of the three dimensions 3, 4, and 5. In general,

The volume of a rectangular solid is the product of its three dimensions.

If we let V represent the volume, then, with the symbols already used for the dimensions,

$$V = l \times w \times d.$$

But $l \times w$ is the area of the bottom or *base* and since $V = (l \times w) \times d$, we can state the rule as follows:

The volume of a rectangular solid is the product of the depth by the area of the base.

The first form of the rule is more suitable for calculations; the second will be useful to us in the considerations of the next article.

If all the dimensions of a rectangular solid are equal, its six faces are all equal squares and the solid is called a *cube*. From the rules for the surface of any rectangular solid and for the area of a square we, therefore, have at once,

The surface area of a cube is six times the square of its edge,

and from the rule for volume,

The volume of a cube is the cube of the length of its edge.

Because of this last rule, volume measure is called cubic measure, and a cube which has an edge of 1 *inch* (centimeter, foot, etc.) is called 1 *cubic inch* (centimeter, foot, etc.).

114. Prisms. A *prism* is a solid two of whose faces are polygons and the rest parallelograms or trapezoids. If the two polygonal faces are equal and parallel and the remaining faces (as many as there are sides to each of the polygons) are rectangles the prism is called a *right prism*. Fig. 26 represents a right prism. In Fig. 26 the parallel polygons are the pentagons *ABCDE* and *FGHIJ*, and the five remaining faces are the rectangles *ABJF*, *BCIJ*, *CDHI*, *DEGH*, and *EAFG*. The polygons are called the ends or *bases* of the prism and the rectangular faces its sides. A polygon may have as few as three sides and any number more. Thus, in Fig. 27, (*a*) represents a right *triangular* prism, having equal triangles *ABC* and *DEF* as bases, and three rectangular faces *ABED*, *BCFE*, *ACFD*.

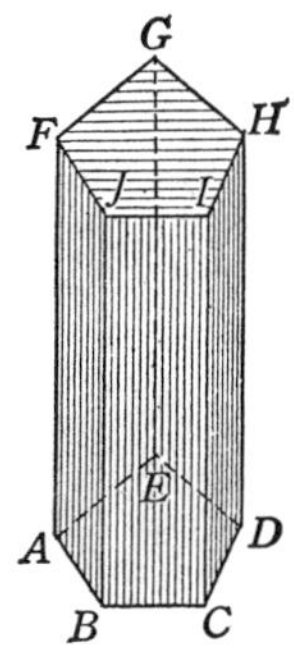

FIG. 26.

As seen in Fig. 27(*b*) a right triangular prism can be thought of as half of a rectangular solid, whose volume is the product of its base area and altitude. But the prism is half the rectangular solid and its base is half the base of the rectangular solid. Therefore, also,

The volume of a right triangular prism is the product of its base area and altitude.

Now, as seen in Fig. 27(*c*), any right prism can be divided into triangular prisms, the sum of whose volumes is, therefore, the volume of the large prism, and the sum of whose bases is the base of the large prism. Therefore,

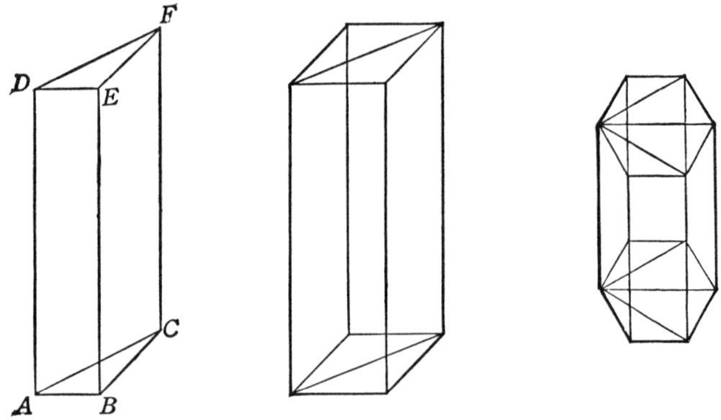

(a) (b) (c)

FIG. 27.

The volume of any right prism is the product of its base area and its altitude.

115. Pyramids. A *pyramid* is a solid which has a polygon as base, and for faces, triangles whose bases are the sides of the polygon and whose vertexes meet in a single point. If the polygonal base is regular and the triangles are all equal, the pyramid is a *regular right pyramid.* Fig. 28 represents a regular pyramid with a pentagonal base, *BCIEF.* The point *A* where the vertexes of the triangular faces meet is called the *vertex* or *apex* of the pyramid and the length of the vertical line *AH* from the vertex to the center of the base is the *altitude* of the pyramid. The altitude *AG* of any one of the triangles forming the faces of the right pyramid is called the *slant height* of the pyramid. The sum of the sides of the base is called the *perimeter* of the base (the distance around it).

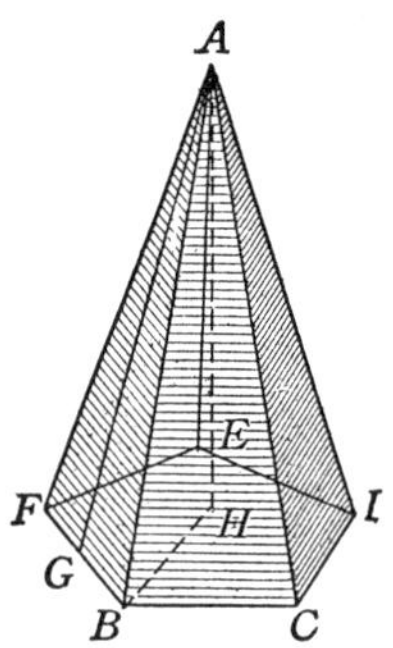

FIG. 28.

The area of any one of the triangular faces is half the product of its altitude and base and the total area of all the faces is the sum of these half products. Since the altitude is the same for all the faces, the total area is half the product of the triangular altitude by the sum of the bases. But the common altitude of the triangular faces is the slant height of the pyramid, the sum of the bases of the triangles is the perimeter of the base of the pyramid, and the sum of the areas of the triangles is the side or *lateral* area of the pyramid. Therefore,

The lateral area of a right pyramid is half the product of its slant height and base perimeter.

In Fig. 29, F is the center point of the cube AB with the square base $BCDE$. By drawing the lines FB, FC, FD, FE to the corners of the base, a regular right pyramid is formed, having the square base $BCDE$ and the vertex F. By drawing lines from F to each of the other four corners, five other such pyramids will be formed, having the other five faces of the cube as bases and the point F as a common vertex. Thus, the cube will be divided into *six* equal pyramids with the base of each equal to a face of the cube and the altitude of each equal to half the depth or height of the cube.

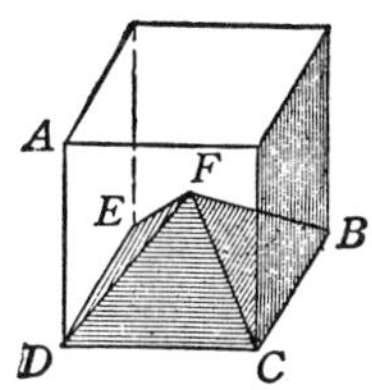

FIG. 29.

Therefore, the volume of each pyramid is $\frac{1}{6}$ the volume of the cube. But the volume of the cube, a rectangular solid, is equal to the product of its height by the area of one of its faces (base). Therefore, the volume of each pyramid is $\frac{1}{6}$ the product of its base area (cube face) by the cube's height, which is *twice* the pyramid's altitude. Therefore,

The volume of the pyramid is one third the product of its base area and altitude.

In solid geometry this is proved to be true for any pyramid.

116. The Regular Solids. A *regular* solid is a solid whose plane faces are all equal regular polygons. These are sometimes called regular *polyhedrons*. It is proved in solid geometry that there are only five such solids. These are the polyhedrons having for faces

(1) Four equilateral triangles (regular pyramid);
(2) Six squares (the cube);
(3) Eight equilateral triangles;
(4) Twelve regular pentagons;
(5) Twenty equilateral triangles.

These five are called, respectively, the regular *tetrahedron*, *hexahedron*, *octahedron*, *dodecahedron*, and *icosahedron*.

These names are formed from the word *hedron*, from the Greek word *edra*, meaning "base," and the Greek prefixes for *four*, *six*, *eight*, *twelve* and *twenty*.

The regular polyhedrons are of importance in higher mathematics, but of little importance in practical measurements and calculations and are described here simply as a matter of interest.

117. Problems.

1. What is the volume in cubic feet of a bin 24 by 48 by 90 inches and how many square feet of lumber is needed to make it, including a cover?
2. The gray limestone of New York state weighs 175 pounds to the cubic foot. What is the weight of a block 1 yard square and 8 feet long?
3. How many bushels will the box of Prob. (1) hold?
4. The Great Pyramid of Egypt has a square base about 750 feet square, is 500 feet high, and the slant height is 625 feet. Find: 1. how many acres of land it covers; 2. how many square yards of lateral surface it has; and 3. if solid, how many cubic yards of stone it contains.
5. What must be the inner edge of a cubical bin to hold 1250 bushels of wheat?
6. If a cubic foot of iron were formed into a bar $\frac{1}{2}$ inch square without waste, how long would the bar be?
7. Find the cost of painting a church steeple of pyramidal form whose base is a hexagon 5 feet on each side, with a slant height of 60 feet, the cost being $2.00 a sq. yd.

B. The Three Round Bodies

118. The Cylinder. If one side of a rectangle is held stationary while the rectangle is rotated about that side as an axis, the other side sweeps out a curved surface which is called a right cylindrical surface or a *right cylinder*. The two ends of the rectangle turn out two equal circles which are called the *bases* of the cylinder. The radius of each of these circular bases is the end of the rectangle and is called the *radius* of the cylinder. The side of the rectangle about which it rotated is now a straight line through the center of the cylinder parallel to the curved surface and is called the *axis* of the cylinder. A cylinder is represented in Fig. 30. In Fig. 30, the line CD is the axis and the distance AB along the axis between the plane circular ends or bases is called the altitude. AP is the radius and the curved surface S is called the *lateral surface* of the cylinder.

Examples of cylinders are an ordinary tin can, a section cut out of a straight round rod, a length of round pipe, a round unsharpened pencil. The length of such a cylinder is the altitude, the circular ends the bases; the diameter of the ends is the *diameter* of the cylinder, and their circumference the *circumference* of the cylinder.

If the curved lateral surface is rolled out straight or flat, as the metal sheet of a cylindrical tin can, or the paper label surrounding such a can, it will form a rectangle, one of whose dimensions is the altitude of the cylinder and the other the circumference of the cylinder. Since the area of this rectangle is the lateral surface of the cylinder and is equal to the product of its dimensions,

The lateral surface of a cylinder is the product of its circumference and altitude.

Now the circumference is, according to article 109, 2π times the radius or π times the diameter; this rule becomes, therefore:

The lateral surface of a cylinder is 6.2832 *times the product of its radius and altitude;* or 3.1416 *times the product of its diameter and altitude.*

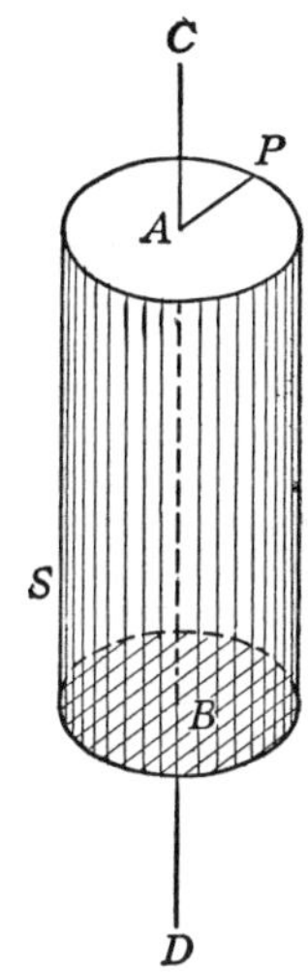

FIG. 30.

Letting S stand for the lateral surface and using other symbols as before, these rules are written insymbols as follows:

$$S = C \times a; \quad S = 2\pi \times R \times a, \quad \text{or} \quad S = \pi \times D \times a.$$

The total surface area consists of the lateral surface and the two circular ends, and is, therefore, the sum of the lateral surface as just found plus twice the area of one of the circles. Using the rule for the area of a circle (article 109), we have

The total surface area of a cylinder is 6.2832 *times the radius times the sum of altitude and radius;* or 3.1416 *times the diameter times the sum of altitude and one half the diameter.*

In symbols these two rules give for the total surface area

$$A = 2\pi \times R \times (a + R); \quad A = \pi \times D \times (a + \tfrac{1}{2}D).$$

If, in Fig. 26, the number of faces of the prism, and number of sides of the base, be increased without at the same time increasing the over-all size of the bases, the prism resembles more and more closely a right circular cylinder. When the number of sides is extremely great and the width of each side or face extremely small, the prism is practically a cylinder. But the volume of the prism is the product of its altitude and base area. Therefore, also,

The volume of a cylinder is the product of its altitude and base area.

But the base area is the area of the circle whose radius or diameter is that of the cylinder. Therefore,

The volume of a cylinder is 3.1416 *times the product of its altitude by the square of the radius;* or, .7854 *times the product of the altitude by the square of the diameter.*

In symbols these rules are written:

$$V = \pi \times a \times R^2, \quad \text{or} \quad V = \frac{\pi}{4} \times a \times D^2.$$

As an illustration of the use of these rules, let us calculate the lateral surface, total surface area and volume of a cylindrical tin can 3 inches in diameter and 10 deep. Here the diameter is $D = 3$ in. and the altitude is $a = 10$ in. The lateral surface is, therefore,

$$S = 3.1416 \times 3 \times 10 = 94.25 \text{ sq. in.};$$

the total surface area is

$$\begin{aligned} A &= 3.1416 \times 3 \times (10 + \tfrac{3}{2}) \\ &= 3.1416 \times 3 \times 11\tfrac{1}{2} = 108.39 \text{ sq. in.}; \end{aligned}$$

and the volume is

$$\begin{aligned} V &= .7854 \times 10 \times 3^2 \\ &= .7854 \times 10 \times 9 = 70.69 \text{ cu. in.} \end{aligned}$$

By working the rules backward, the diameter or radius can be found when the altitude and the surface or volume are known, and the altitude can be found when the diameter or radius and the surface or volume are known.

119. The Cone. If one leg of a right triangle is held stationary while the triangle is rotated about that leg as an axis, the hypotenuse sweeps out a curved surface which is called a conical surface or a *right circular cone.* The other leg of the triangle turns out a circle which is called the *base* of the cone, whose radius is the length of the rotating leg and is called the base *radius* of the cone. The curved conical surface is called the *lateral surface* of the cone. Fig. 31 represents a cone formed in this manner with the right triangle *ABC*. The line *DE* is the direction of the stationary leg *AC* and is called the axis of the cone. The point *A* is the apex or *vertex* of the cone, the circle with center at *C* is the base, *CB* is the base radius, *AC* is the *altitude*, and *AB* is the *slant height.*

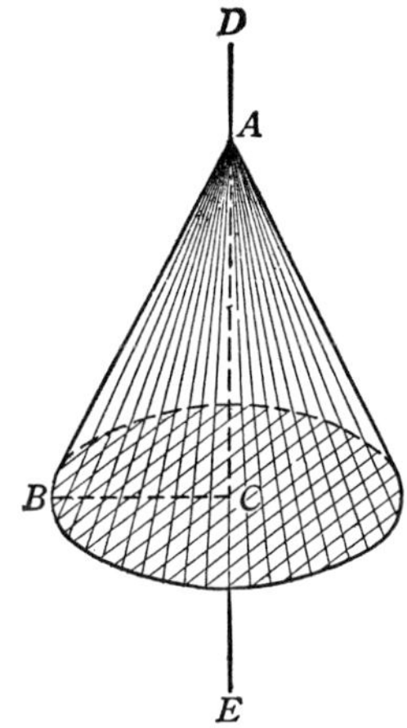

FIG. 31.

Examples of cones are the round, pointed end of a sharpened pencil; round, pointed paper cups; a round, tapering pointed lamp shade; etc.

If, in Fig. 28, the number of faces of the pyramid, and number of sides of the base, be increased without at the same time increasing the over-all size of the base, the pyramid resembles a cone more and more closely. When the number of sides and faces is extremely great and the width of each side or face is extremely small the pyramid is practically a cone, its base perimeter becomes the circumference of the base of the cone, and its slant height, lateral surface and volume become the slant height, lateral surface and volume of the cone. By the rule for the lateral surface of the pyramid, therefore,

The lateral surface of a cone is half the product of its base circumference and slant height.

Now the slant height *AB* in Fig. 31 is the hypotenuse of the right triangle *ABC*, one leg of which is the altitude *AC* and the other the radius *BC*. From the hypotenuse rule for right triangles, therefore,

The slant height of a right cone is the square root of the sum of the squares of its altitude and base radius.

Also, the circumference of the base, according to the circle rule, is 2π times the radius. Using this relation and the rule just found for slant height, the rule given above for the lateral surface of the cone becomes:

The lateral surface of a right cone is 3.1416 *times the product of its radius and slant height.*

Letting s represent slant height and using other symbols as for the circle and cylinder these rules become, in symbols:

$$s = \sqrt{R^2 + a^2}; \quad \text{and} \quad S = \pi \times R \times s.$$

The total surface area is the sum of the lateral surface and the area of the circular base. Combining the rules for the lateral surface and the area of a circle, therefore, we have

The total surface area of a right cone is 3.1416 *times the radius of the base times the sum of the radius and the slant height.*

In symbols this is

$$A = \pi \times R \times (R + s).$$

In case the base diameter is known, half the diameter is the radius to be used in the rules for the cone.

Next consider the volume: since the volume of the cone is found from that of the pyramid, as seen above, we have, therefore,

The volume of a cone is one third the product of its altitude and base area;

and since the base is a circle,

The volume of a cone is 1.0472 *times the altitude times the square of the radius;* or .2618 *times the altitude times the square of the diameter.*

In symbols this is

$$V = \frac{\pi}{3} \times a \times R^2, \quad \text{or} \quad V = \frac{\pi}{12} \times a \times D^2.$$

As an example of the application of the various rules for the cone let us find the slant height, lateral and total surface, and the volume of a cone 10 inches high and 6 inches across the base. Here the base diameter is 6 and, therefore, the radius is $R = 3$ inches, and the altitude is $a = 10$ inches. By the rules, therefore, the slant height is

$$s = \sqrt{(3^2 + 10^2)} = \sqrt{109} = 10.44 \text{ in.};$$

the lateral surface is

$$S = 3.1416 \times 3 \times 10.44 = 98.39 \text{ sq. in.};$$

the total surface area is

$$\begin{aligned} A &= 3.1416 \times 3 \times (3 + 10.44) \\ &= 3.1416 \times 3 \times 13.44 = 126.67 \text{ sq. in.}; \end{aligned}$$

and the volume is

$$\begin{aligned} V &= 1.0472 \times 10 \times 3^2 \\ &= 1.0472 \times 10 \times 9 = 94.25 \text{ cu. in.} \end{aligned}$$

120. The Sphere. A *sphere* is a globe, a perfectly round ball; a very familiar object and in form the simplest of the solids, just as the circle is in form the simplest of plane figures. It has vied with the circle in historical interest and has, in spite of its apparent simplicity, many very interesting properties.

The *center* of a sphere is the middle point of its volume, the one point equally distant from every point of its surface. Any line from the center to the surface, or the length of such a line, is a *radius* of the sphere. Any line through the center, with its two ends in the surface, or the length of such a line, is a *diameter* of the sphere. The diameter is, therefore, twice the radius, or the radius is half the diameter, as in the case of the circle.

The sphere may be formed by means of the circle, and its properties determined from those of the circle. Thus, if one diameter of a circle be held stationary while the circle is rotated about it as an axis, the circumference of the circle sweeps out the surface of a sphere which has the same center, radius and diameter as the circle. This circle is called a *great circle* of the sphere and the circumference of the great circle is the *circumference* of the sphere. *Any* circle having the same diameter and circumference as the sphere is also called a great circle. If a sphere is cut straight through the center, it is divided into two equal parts and each is called a *hemisphere.* If a great circle is drawn around the sphere on its surface (as the equator of the earth) the surface of the sphere is also said to be divided into two hemispheres.

If a sphere rotates about a diameter as an *axis*, the two ends of that diameter are called the poles of the rotating sphere and the circumference of a great circle whose plane is perpendicular to the axis at the center, that is, a line around the surface of the sphere midway between the poles, is called the *equator.* Any great circle passing through both poles is called a *meridian circle.*

Some of the definitions and dimensions which we have just given concerning the sphere are the same as those used in the treatment of latitude and longitude in Chapter 15. As used in that chapter, the names and meanings are derived from the sphere as here described; they are applied to the earth because it is very nearly an exact sphere.

We next proceed to study the properties of the sphere and their calculation.

If a hard smooth string or thin cord is smoothly and closely wound over the curved surface of a hemisphere, and also over the surface of its great circle, so as to just cover both, as in Fig. 32(*a*) and (*b*), re-

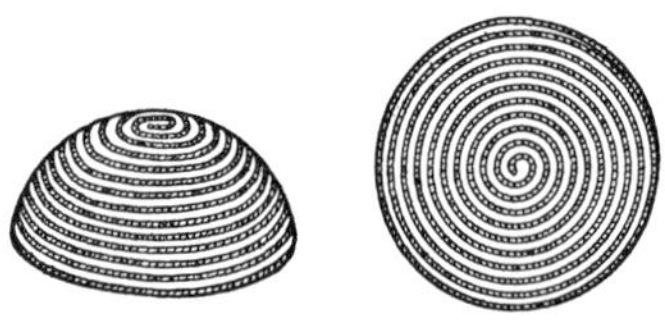

(*a*) (*b*)

FIG. 32.

FIG. 33.

spectively, it is found that just twice as much cord is required to cover the hemisphere as is required for the great circle. The curved surface area of the hemisphere is, therefore, just twice the area of the great circle. Therefore,

The surface area of a sphere is four times the area of its great circle.

If a whole sphere is divided into a large number of small pyramid-shaped sections, as indicated in Fig. 33, so that the vertex of each pyramid is at the center of the sphere, then the altitude of each small pyramid is the radius of the sphere, the sum of the base areas of the pyramids is the surface area of the sphere, and the sum of the *volumes* of the pyramids is the *volume* of the sphere. According to the rule for the volume of a pyramid, therefore, the volume of the sphere is one third the sum of the products of the pyramidal bases by the common altitude, or one third the product of the *common altitude* by the *sum of the bases*. That is,

The volume of a sphere is one third the product of its radius and surface area.

But the surface area is given by the first rule stated above. Using that rule, therefore, the last rule becomes

The volume of a sphere is four thirds the radius times the area of its great circle.

Thus, the first and third of these rules give the surface area and volume of a sphere in terms of the area of its great circle. Both these rules are strictly and exactly proved in solid geometry.

Now the area of the great circle is given by the circle rule when the radius or diameter is known and these are the same as the radius and diameter of the sphere. Using the diameter, therefore, the first and third rules above become:

> *The surface area of a sphere is* 3.1416 *times the square of its diameter.*
> *The volume of a sphere is* .5236 *times the cube of its diameter.*

Using abbreviations as before, these rules become, in symbols:

$$A = \pi \times D^2; \quad V = \frac{\pi}{6} \times D^3.$$

The surface area and volume of a 10 inch ball, or sphere, are therefore, for example,

$$A = 3.1416 \times 10^2 = 3.1416 \times 100 = 314.16 \text{ sq. in.};$$
$$V = .5236 \times 10^3 = .5236 \times 1000 = 523.6 \text{ cu. in.}$$

121. Problems.

1. Find the lateral area of a cylinder of altitude 2 ft. 5 in. and base circumference 4 ft. 9 in.
2. What is the total surface area of a cylinder formed by the rotation about one of its sides of a rectangle that is $6\frac{1}{2}$ feet long and 4 feet wide?
3. Find the volume of a cylinder of altitude 15 feet and radius 1 ft. 3 in. ($1\frac{1}{4}$ or 1.25 feet.)
4. Find the lateral surface of a cone of base diameter $17\frac{1}{2}$ feet and slant height 30 feet.
5. Find the total surface area of a cone of base diameter $6\frac{3}{4}$ feet and slant height 45 feet.
6. Find the volume of a cone of altitude 24 feet and base diameter 30 inches.
7. What is the surface area of a 9 inch sphere?
8. What is the surface area of a sphere whose circumference is 31.416 inches?
9. Taking 1° of longitude at the equator as 69.2 miles and the earth as a sphere, find its diameter in miles, its surface in sq. mi., and its volume in cu. mi.
10. How many gallons will a cylindrical tank hold, being 10 feet in diameter and 10 feet deep?
11. If a cylindrical bushel measure is to be 10 inches deep, the curved side being made of a single piece of sheet metal 10 inches wide, and if 1 inch is allowed for lapping and riveting, how long must the piece be?
12. A spherical iron cannon ball of .271 pound to the cubic inch weighs 117 pounds. What is its diameter?
13. How many four-inch globes are equivalent in volume to a one-foot globe?
14. How many bushels of wheat are contained in a conical pile 8 feet across the base and 8 feet high?

15. A conical paper cup is 4 in. deep and 3 in. across. How many such cups are required to hold 2 gal., and how much paper is required to make them?

16. A spherical toy balloon has air blown in to increase its diameter from 5 to 6 inches. How much air is blown in?

17. A ball 2 inches in diameter just fits snugly inside a square (cubical) box. How much empty space is there inside the box around the ball?

18. Canned fruit in cans 3 inches in diameter and 5 inches deep is packed in a box having inside dimensions of 5 x 12 x 18 inches. How much waste space is there inside the box between the cans?

19. A gasoline tank has the shape of a cylinder with hemispherical ends. If the diameter of the cylindrical part is 4 feet and its total length is 14 feet how many gallons will it hold when entirely full?

20. A large drill bit is $1\frac{1}{2}$ inches in diameter and the end (point) is ground in conical shape so that the drill is $\frac{1}{2}$ inch longer at the center than at the outside. When it has drilled a hole which is $3\frac{1}{2}$ inches deep at the center, how much metal has been removed from the hole?

Part 4

SOME SPECIAL APPLICATIONS

Chapter 18

GRAPHS

122. Data, Statistics and Graphs. A fact or an item of definite information which is to be used as a basis for comparison, discussion or conclusions is called a *datum* (plural, *data*). When several data or a large number are systematically collected and arranged, to be used for reference and study, they are called *statistics.*

Data and statistics are usually expressed in terms of numbers arranged or classified in sets which are related in a definite way to one another or to some system of reference. Thus, a list of the amounts of rainfall in inches or the average temperature in degrees corresponding to the months of the year; a list showing the number of inhabitants of a certain territory or country for each of a number of years or census periods; census information in general; the prices of stocks or bonds for each day, week, month or year over definite periods; these are illustrations of statistics.

There are several methods of recording and presenting statistics. A very common method is that of listing the numbers in tables which are divided into columns, each column being numbered or headed with some explanatory title or classification. Thus the first example given above might be listed or tabulated in three columns: one containing the names of the months of the year from January to December in order, and the other two the numbers showing the amounts of rainfall and average temperatures, respectively, corresponding to each of the months. This method of presenting data or statistics is referred to as *tabulation* and is much used in business, industry and science.

In order to study and compare tabulated data or statistics it is necessary to examine each individual item or number separately. This method of study is slow and tedious and does not give a simple and complete mental picture of the whole body of data and their relations and variations under varying conditions. This result is best accomplished by a pictorial or *graphical* representation of the data. Such pictures are called *graphs.*

Thus, if a picture of a row of tall slender glass tubes, one for each month, and each filled to a depth which represents the amount of rainfall for that month, is used, a single glance at the picture (or graph) is sufficient to show the separate amounts of rainfall, the comparison for the different months, and the changes throughout the year.

A similar picture or graph might be drawn to represent a thermometer tube for each of the months with the height of the mercury (quicksilver) in each tube representing the average temperature for that month. This graphical representation of the data is more compact and more easily studied and remembered, and also allows estimates to be made of values or numbers not given in the tabulated data. On this account and for various reasons which will appear as we proceed, different forms of graphs are much used in the recording and study of data and statistics.

123. Forms and Meanings of Graphs. Suppose that for a certain year in a certain neighborhood the weather bureau records of monthly

Month		Rainfall (Inches)	Average Temperature (Deg. F.)
No.	Name		
1	January	3.3	20
2	February	4.5	22
3	March	3.8	28
4	April	3.9	35
5	May	4.1	45
6	June	3.3	65
7	July	1.5	80
8	August	2.5	85
9	September	6.5	75
10	October	3.8	60
11	November	5.0	35
12	December	3.9	25

rainfall and monthly average temperature gives the values shown in the table above. These data are represented by the graphs of Fig. 34, (*a*) showing the monthly rainfall in inches and (*b*) the monthly average temperature in degrees Fahrenheit.

In Fig. 34(*a*) the length of each of the vertical black lines represents the depth in inches of the rainfall for the month shown at the foot of that line, $\frac{1}{8}$ inch on each line representing 1 inch of rainfall, and the amount being indicated on each line by the number taken from the table. In Fig. 34(*b*) the lengths of the vertical lines similarly represent the height of the thermometer reading above zero degrees Fahrenheit, $\frac{1}{5}$ inch on each line representing 10 degrees.

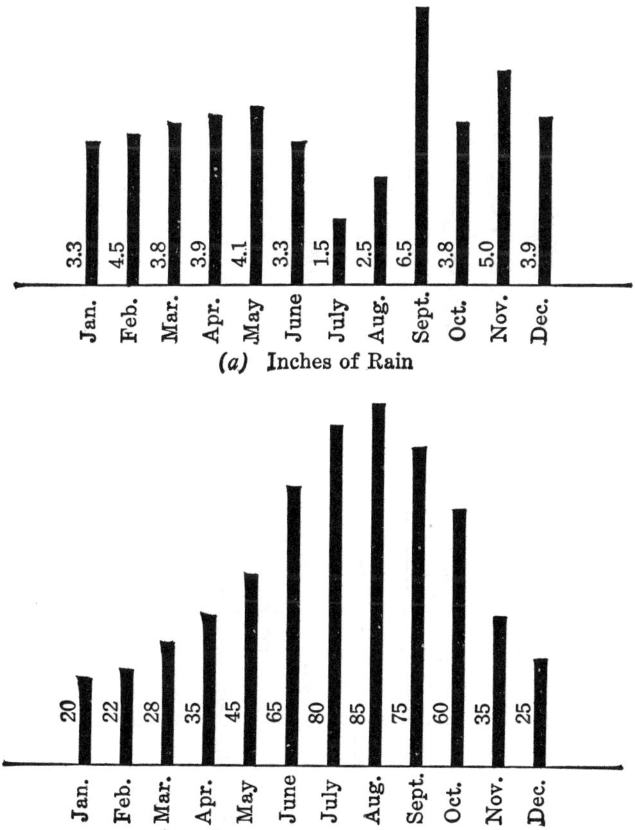

(*a*) Inches of Rain

(*b*) Degrees of Temperature

FIG. 34.

The convenience of these graphs is at once evident; the amount of rainfall and the intensity of heat for each month, the comparative values for different months, the total range of values (highest and lowest) for the year, and the variations during the year being seen at a glance.

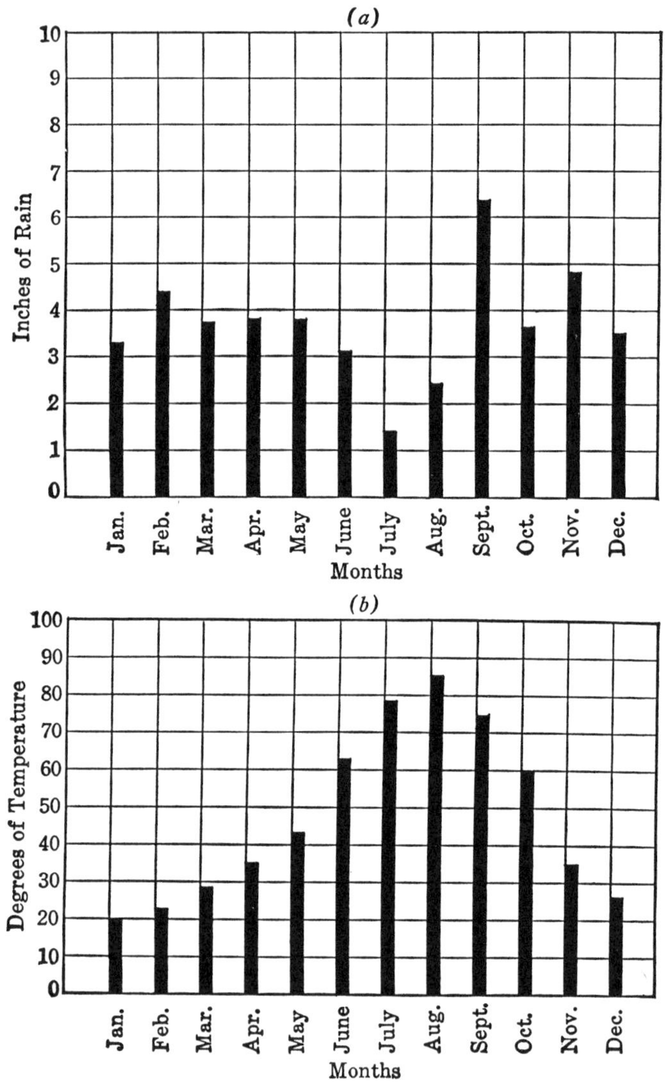
(a)
10
9
8
7
6
5
4
3
2
1
0
Inches of Rain
Jan.
Feb.
Mar.
Apr.
May
June
July
Aug.
Sept.
Oct.
Nov.
Dec.
Months
(b)
100
90
80
70
60
50
40
30
20
10
0
Degrees of Temperature
Jan.
Feb.
Mar.
Apr.
May
June
July
Aug.
Sept.
Oct.
Nov.
Dec.
Months

FIG. 35.

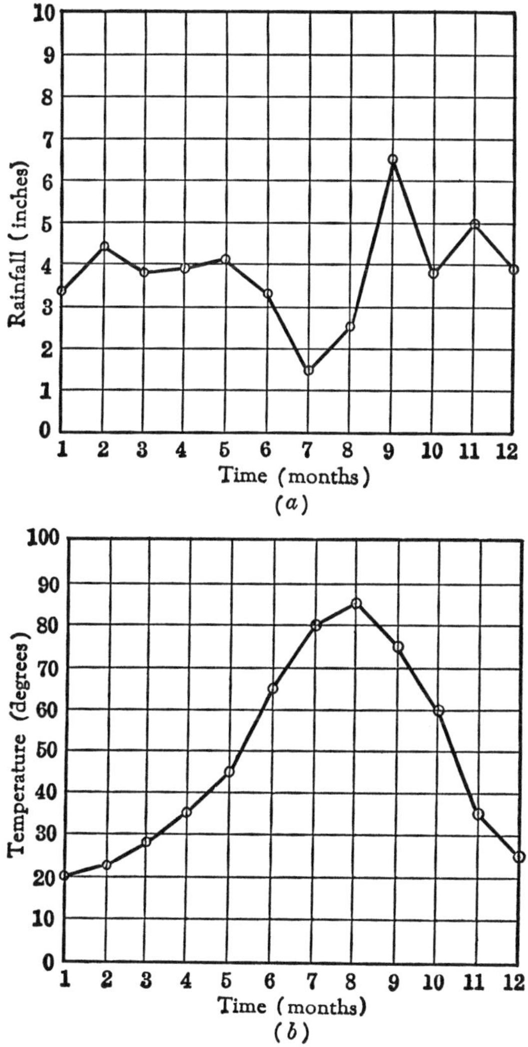
10
9
8
7
6
5
4
3
2
1
0
Rainfall (inches)
1 2 3 4 5 6 7 8 9 10 11 12
Time (months)
(a)
100
90
80
70
60
50
40
30
20
10
0
Temperature (degrees)
1 2 3 4 5 6 7 8 9 10 11 12
Time (months)
(b)

Fig. 36.

By using vertical lines of two different colors, say black for rainfall and red for temperature, both sets of data could be represented on one sheet, using only one base line on which to erect the verticals. This method is not so convenient, however, as is a method which will be described later.

The graphs of Fig. 34 are shown in a slightly different form in Fig. 35, in which (*a*) and (*b*) each represents the same data as before. Instead of showing the figures for each month on the corresponding vertical line, they are all shown on a single vertical line at the left and the horizontal lines running across the graph enable the figures corresponding to any month to be read at once. The base or reference line showing the months or time and the reference line at the left showing the rainfall or temperature are called the axes of the graph, the one being called the *time axis* and the other the *rainfall* or *temperature axis.* Both these methods (Fig. 34 and Fig. 35) are very much used in graphical presentation of data and statistics.

If, in Fig. 35, lines be drawn joining the upper end of each vertical line to that of the next and the heavy vertical line itself be omitted (only its end-point being marked by a little circle) the graphs of Fig. 36 result. Here the rainfall and temperature axes are the same as before, but the time axis shows the months by number instead of by name, the numbers being more easily and simply written and printed. The graph in this form (for each case) is called a *curve*, although more usually a curve is not made up of short sections of straight lines, but consists of one smoothly curving line.

By the use of *curves*, both sets of data are easily represented on one *graph sheet*, as already mentioned. This is done in Fig. 37. Here the time axis is the same for both curves, the rainfall axis is at the left and the temperature axis at the right. Both curves are shown as smooth curves, the rainfall curve being dotted and the temperature curve solid, and the two being indicated by (*a*) and (*b*) as before.

The various forms of graphs shown in Figs. 34 to 37 illustrate the ones most used of the different possible forms. The paper may be blank paper with the lines drawn in for each graph as the graph is *constructed*, as in Fig. 34, or it may be *ruled paper*, paper printed with the lines and squares already prepared, such as that used in Figs. 35–37. Ruled paper is, however, generally used and it may be ruled in any desired manner, as shown in some of the graphs of article 125.

Different curves may require different axes, as in Fig. 37, or they may be drawn with the same axes as in Fig. 42 (article 125). If two or more curves are to be drawn on one sheet they may be drawn in a different manner (solid, dotted, dot-and-dash, etc.) as in Fig. 37, or they may be drawn solid but in different colored inks.

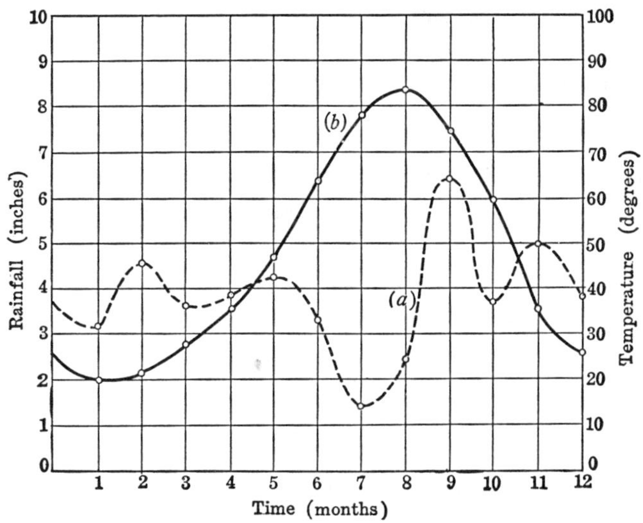

FIG. 37.

In this article one set of data has been used to show several different forms of graphs and the meaning of each, but obviously many different kinds of data and statistics may be represented by graphs of different forms. This fact is illustrated in article 125.

124. Curves and Curve Plotting. There are several terms used in connection with such graphs as those of Fig. 37 which it is well to know.

The measured quantities or numbers which are represented or laid off on the horizontal axis are called *abscissas* and those represented or laid off on the vertical axis are called *ordinates*. One of the small circles in Fig. 37, which marks the upper end of an ordinate line drawn at the end of a certain abscissa which is measured from the point O, is called a *point on the curve*.

In laying off the lengths of the abscissa and ordinate to locate such a point we are said to *plot* the point, and the abscissa and ordinate together for each point are called the *coordinates* of the point. In plotting all the points whose coordinates are known and then drawing the curve to join all these points we are said to *plot the curve* or the graph representing the data which form the coordinates.

The paper which is ruled off in squares for the plotting of curves is called graph paper, or preferably *coordinate paper.* Coordinate paper may be purchased which is of any desired size and ruled in any desired manner.

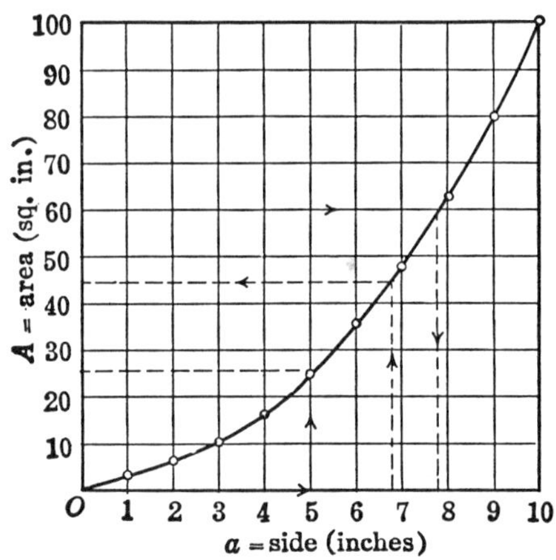

FIG. 38.

Curves can be plotted to represent the relation between any two related quantities or measurements, and not only statistics. Thus let us consider the areas of squares whose sides have different lengths. If a represents the length of the side in any convenient measure (inches, centimeters, feet, etc.) then the area A in the corresponding square measure is the square of the side. That is, $A = a^2$. If the sides have the values 1, 2, 3, 4, etc., then the corresponding values of the areas A are 1, 4, 9, 16, etc. Tabulating a few of these values we get the table shown below at the right. Of course, if the side is zero the area is zero; that is, there is no square.

By plotting a curve with values of the sides (a) as abscissas and values of the areas (A) as ordinates we get the curve of Fig. 38. The manner of plotting the points on this curve will be illustrated if we consider the coordinates of the point corresponding to the side $a = 5$. From the table these coordinates are 5 and 25. With the abscissas a from 0 to 10 and the ordinates A from 0 to 100 already marked off on the axes, we go out from the zero point 5 units on the axis of abscissas, and from this point 25 units up on the vertical line, as indicated by the dotted line drawn to the axis of ordinates. At the point thus reached draw the small circle. The other points whose coordinates are given are plotted in the same manner and the curve is drawn through them. The curve can, of course, be extended as far as desired by using a larger sheet and plotting more points.

Side	Area
a (inches)	$A = a^2$ (sq. in.)
0	0
1	1
2	4
3	9
4	16
5	25
6	36
7	49
8	64
9	81
10	100
etc.	etc.

The curve can now be used to read off without calculation the area of a square of any side (within the limits of the curve). Thus for $a = 6.7$: if we locate 6.7 on the abscissa axis, go upward from this point to the curve as shown by the dotted line, go to the left to the axis of ordinates as shown by the horizontal dotted line; this axis is reached at 45. That is, for a side $a = 6.7$ inches, the area is $A = 45$ sq. in.

Again, suppose the area of a certain square is 60 sq. in.; what is the side? Locating 60 on the axis of ordinates and following the horizontal line to where it meets the curve as shown by the arrow, and going downward from this point as shown by the dotted line, the axis of abscissas is reached at $7\frac{3}{4}$ or 7.75. That is, for an area $A = 60$ sq. in., the side of the square is $a = 7.75$ or $7\frac{3}{4}$ inches.

In using the curve as illustrated in each of these examples it is not necessary to draw such dotted lines as those shown here for explanation. Very closely ruled paper allows the values to be read directly, or a straightedge or ruler may be laid on the graph instead of drawing the lines.

Curves in great variety are plotted and used extensively in the manner here illustrated, in the higher mathematics and in science and engineering.

125. Examples of Graphs. We give here a few further examples illustrating the construction of graphs and plotting of curves, without explanations of their uses. Brief explanations of their meanings are given wherever they seem necessary.

Distances of Planets from the Sun (Fig. 39). This is a very simple graph showing more expressively than the numbers the comparisons of the distances of the planets from the sun.

DISTANCES FROM THE SUN

Planet	Millions of Miles	Planet	Millions of Miles
Mercury	36	Jupiter	481
Venus	67	Saturn	886
Earth	93	Uranus	1782
Mars	142	Neptune	2778

FIG. 39.

There is a very small planet (about 100 miles in diameter) between Mars and Jupiter at a distance of 257 millions, and a recently discovered planet beyond Neptune at an estimated distance of about 3675 million miles. This planet has been named Pluto.

Growth of A Manufacturing Corporation. The Star Manufacturing Corporation manufactures shoes. The graphs on page 195 show the growth in the gross value of the wholesale sales of this corporation during the years 1957–1962 and in the net profits during the same years (in round figures).

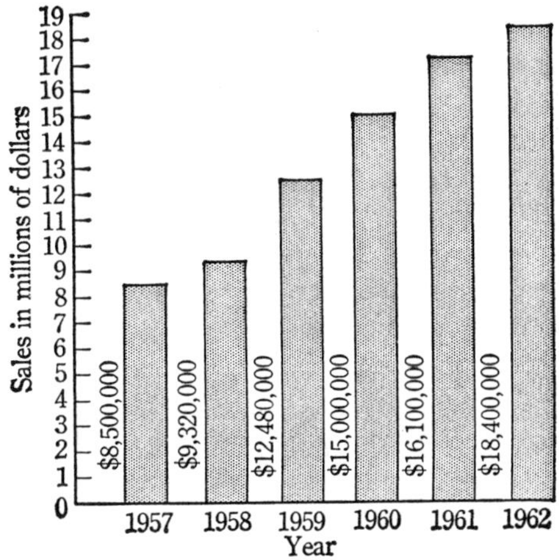
Sales in millions of dollars
19
18
17
16
15
14
13
12
11
10
9
8
7
6
5
4
3
2
1
0
$8,500,000
$9,320,000
$12,480,000
$15,000,000
$16,100,000
$18,400,000
1957
1958
1959
1960
1961
1962
Year

FIG. 40.

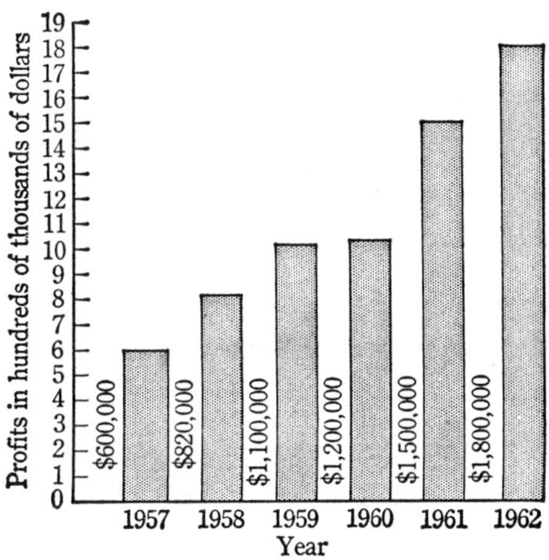
Profits in hundreds of thousands of dollars
19
18
17
16
15
14
13
12
11
10
9
8
7
6
5
4
3
2
1
0
$600,000
$820,000
$1,100,000
$1,200,000
$1,500,000
$1,800,000
1957
1958
1959
1960
1961
1962
Year

FIG. 41.

POPULATION OF THE TOWN OF SPRINGFIELD

Year	Total Population	School Population
1940	58,136	8,112
1942	53,208	7,426
1944	48,847	6,908
1946	55,329	7,814
1948	56,012	9,003
1950	57,429	9,125
1952	62,138	9,340
1954	63,896	10,093
1956	65,552	10,126
1958	69,005	11,430
1960	70,150	13,257
1962	75,210	14,786

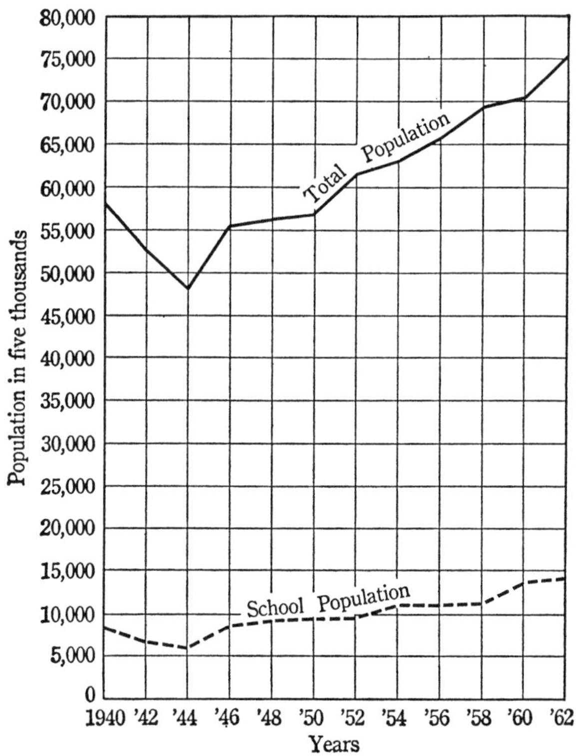

FIG. 42.

Population of the Town of Springfield. In the table on page 196 are listed the population of the town of Springfield and the school population of the same town for the years listed.

On the graph, the total population is indicated by the solid line and the school population by the dotted line. A graph of this type, used with caution, may be used to estimate future school population and the need for new school facilities.

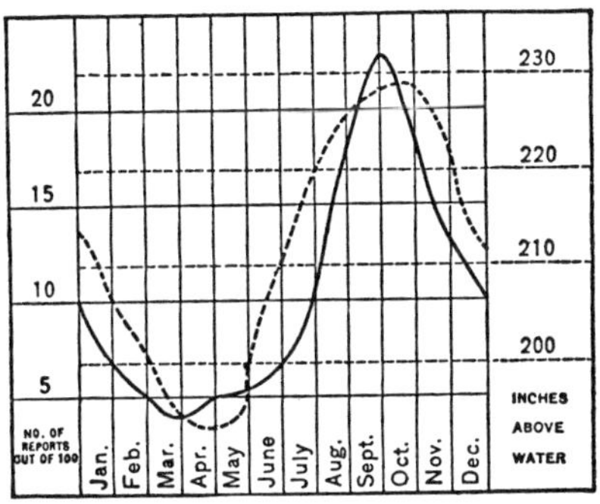

FIG. 43.

Relation of Typhoid Prevalence to Water Supply Conditions (Fig. 43). In Fig. 43 the dotted curve shows the average distance from the surface of the ground to the surface of water in the wells of a certain state for the months of a certain year, as determined by the board of health of that state. When this curve is the *highest* the water in the wells is the *lowest.* The ordinates for this curve are at the right.

The solid line curve shows the number of reports in every hundred in which typhoid fever was mentioned as prevalent, and applies to the same territory for the same time as the dotted curve. The ordinates for this curve are at the left. The months are shown at the bottom for both curves.

The correlation between low water and typhoid prevalence is to be noted. The two conditions occur together throughout the year. The conclusion is that low water is an important contributing cause to typhoid fever.

It is also seen that the healthiest season of the year, as far as typhoid fever is concerned, was the spring (April and May) and the most dangerous the early fall (September and October).

Surface of Cylinders (Fig. 44). In the table below are given the total surface areas (S) of cylinders of different radii (R) up to 10 inches (20 inches diameter) when all have the same height (altitude) of 10 inches. The total surface area is calculated to the nearest whole square inch for each case by the rule of article 118, which, in symbols, is

$$S = 6.283 \times R \times (R + 10).$$

The results are plotted in Fig. 44 with the radius R in inches as abscissa and the area S in square inches as ordinates.

The curve of Fig. 44 can be used in the same manner as that described in connection with Fig. 38. Thus, from the curve, a cylinder of radius $R = 4\frac{1}{2}$ inches (height 10 inches) has a total surface $S = 400$ sq. in. Similarly, 800 sq. in. of sheet metal (if properly cut) will make a closed cylindrical tank 10 inches deep and 14.8 inches diameter (7.4 in. radius).

Radius	Total Surface
R (in.)	$S = 2\pi R(R + 10)$ (sq. in.)
0	0
1	69
2	151
3	245
4	352
5	472
6	605
7	748
8	907
9	1075
10	1257

This curve (Fig. 44) differs less from a straight line than any we have considered so far. Many graphs are exactly straight lines. The general name *curve* is applied to all, however, whether actually curved or straight.

126. Problem Solution by Graphs. In addition to presenting statistics in a form convenient for study and comparison, graphs may also frequently aid in the solution of certain problems, as illustrated in connection with Figs. 38 and 44. These illustrations make use of curves, but a graph may be *any* form of diagram which will serve to present data or show the relation between related numbers.

A problem solution which is obtained by means of a graph of any kind is called a *graphical solution.*

In this article we give as illustrations the graphical solutions of two problems, one of which is applicable to many problems of a certain general type.

Problem 1. Two trains leave the same place at different times, traveling in the same direction, one at 20 and the other at 40 miles an hour. If the second (faster) train leaves three hours after the first, when and where will it pass the first?

Solution. In Fig. 45 let each space along the abscissa axis OX represent 1 hour, and each space along the ordinate axis OY 20 miles. Then at the end of 1 hour the first train has traveled 20 miles and the

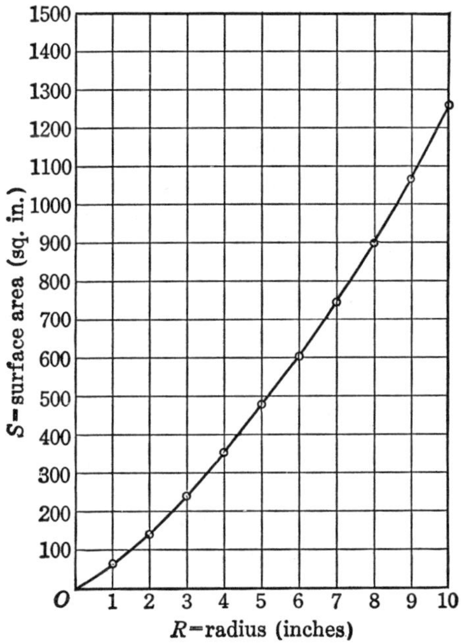

FIG. 44.

point A represents its time and distance from the starting point O. Similarly, after 2 hours, or 40 miles, B represents this relation again, and so on, for C, etc.

As the second train is still at the station (zero distance) at the end of three hours, O' represents its start. Then after 1 hour from its start, or 4 hours from O, it is 40 miles from O and A' represents its position. Similarly, after 2 hours (40 miles farther) B' represents its position, and so on.

Drawing a line through all the set of points representing each train, and continuing these as far as desired, the time-distance graph for each train is obtained. As seen, both are straight lines.

The object of the problem is to find the point on the graph which indicates that both trains are at the same place at the same time. This one point must, therefore, be on *both* lines, that is, *where they meet.*

Extending the graphs it is seen that they meet at the point P, which represents the time 6 hours and 120 miles from O. This is 6 hours after the first train started and 3 after the start of the second, and 120 miles from the starting point.

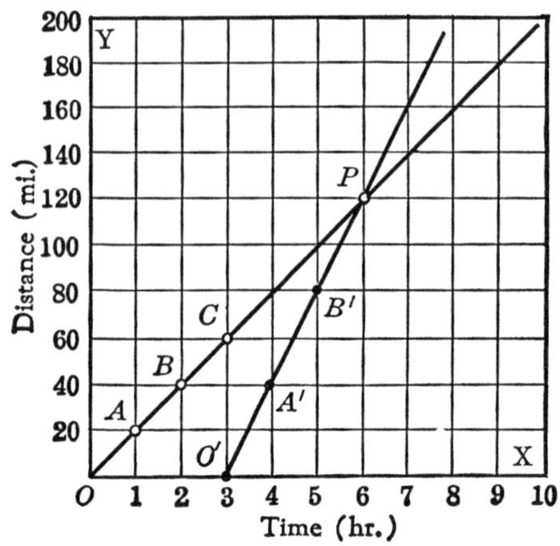

FIG. 45.

The same graph can be used to find how far apart the trains are at any time after the second train started.

Problem 2. A can do a job in 6 days and B can do it in 4 days. How long will it take both, working together, to do it?

Solution. On the line OX in Fig. 46 and at any convenient distance AB apart draw the two parallel lines AC and BD of such lengths that one, say AC, represents the time in which A can do the job, and the other the time in which B can do it, letting the same length represent 1 hour on both.

Next draw the diagonals AD and BC, meeting at the point E; and then draw the line EF parallel to AC and BD.

The length of EF represents the time in which A and B together can do the job, and on being measured it is found to be $2\frac{2}{5}$ days.

This graph may be used to solve any problem similar to the one just solved, such as the following:

Each of two pipes alone can fill (or empty) a tank in a specified time; how long will be required for both running together to fill (or empty) it?

The resistance of each of two electrical conductors is known; what is their joint resistance in parallel?

The capacity of each of two electrical condensers is known; what is their combined capacity in series?

These represent a class of problems which are called "joint effect" problems, and all these and many more can be solved by the graph of Fig. 46 in the same manner: Simply lay off the base line and on it

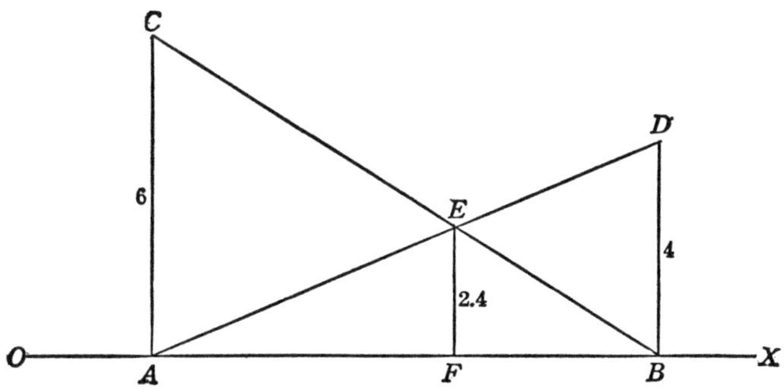

FIG. 46.

erect (to proper scale) the parallel lines *AC* and *BC* representing the "effect" of each of the two actions or things considered; draw the diagonals meeting at *E*, and the line *EF* parallel to *AC* and *BD*; this line then gives the "joint effect" on the same vertical length scale as *AC* and *BD*. If ruled paper is used, the laying off and reading of the lengths are very simple.

If three effects are to be combined, first find the joint effect of any two, then in the same way combine this result with the third; and so on for any number.

127. Exercises and Problems.

1. From Fig. 38 determine the side of a square whose area is 75 sq. in.
2. From Fig. 44 find the amount of sheet tin required to make a cylindrical can 10 in. deep and of 5 in. diameter.

3. The following table shows the population for the cities of a certain state. Prepare a graph of the form of Fig. 40 for this population data.

Dover	35,627	Canton	41,580
Greenville	47,019	Walton	59,237
Bolton	26,122	Chester	18,985
Ashville	30,018	Thorne	52,692

4. From Fig. 40 determine between which two years the greatest growth in sales took place.

5. From Fig. 41 determine between which two years the smallest growth in profits took place.

6. From Fig. 42 estimate the total population in 1953.

7. Prepare for your own state a population graph similar to Fig. 42, from the time of the earliest records to the latest, and predict its probable population 10 years hence.

8. A certain pipe alone can fill a tank with water in 3 hours and another pipe alone in 5 hours. By means of a graph like that of Fig. 46 determine the time required by both pipes together to fill the same tank.

9. Determine from Fig. 45 how far apart the trains referred to in that graph will be 4 hours after the second train starts.

10. At what times will the trains of Fig. 45 be 40 miles apart?

Chapter 19

PERCENTAGE

128. Introduction. In this chapter and the next we discuss a few special but common commercial applications of arithmetic. It is not the purpose of this book to treat generally of technical, industrial and commercial applications of arithmetic, but the percentage method of expressing fractional parts is so widely used, not only in business but elsewhere also, and the use and calculation of interest are of such importance that two short chapters will be devoted to brief descriptive treatments of these subjects.

129. Meaning of Percentage. The words *per cent.* are a contraction of the Latin phrase *per centum* and mean "by the hundred," that is, a certain part of every hundred of any thing or denomination. Thus, 4 per cent. means 4 of every hundred and may signify 4 cents of every 100 cents, 4 dollars of every 100 dollars, 4 persons of every 100 persons, 4 pounds of every 100 pounds, etc. This may be written $\frac{4}{100} = \frac{1}{25}$; similarly 65 per cent., or as we shall write it, 65 percent, means 65 in 100, or $\frac{65}{100} = \frac{13}{20}$; etc. Any percent is therefore a fraction and since the denominator is always 100 it is properly expressed as a decimal fraction. Thus, 4 percent $= \frac{4}{100} = .04$; 65 percent $= .65$, $12\frac{1}{2}$ percent $= .125$; etc.

When a fractional *part* of a whole or total is expressed in hundredths or percent it is referred to as a *percentage* of the total. The word "percentage" is also used to refer to the method or general system of calculation percent, by the hundred, in any branch of arithmetic. Thus, if half of a job is finished, it is 50 percent finished; if any quantity increases by $\frac{1}{8}$ of its value, it has grown $12\frac{1}{2}$ percent; if 4¢ per dollar is lost in a transaction involving any number of dollars, it is referred to as a 4 percent loss; etc. The basis of reckoning is 1 or the total, taken as 100 percent, and the fractional parts are each said to be a certain percentage.

The words "per cent." or simply *percent* are usually represented by the symbol %, which is derived from the fractional form, as $\frac{40}{100} = 40\%$. Thus, 4 percent is written as 4%; $\frac{13}{20} = \frac{65}{100} = .65 = 65\%$; etc.

130. Fractions and Ratios Expressed as Percentage. Any common fraction may be expressed as a decimal fraction (article 34) and any decimal fraction may be expressed as hundredths; therefore, any common or decimal fraction may be expressed as a percentage. As a ratio is simply a common fraction whose numerator and denominator represent the same kind of quantity any ratio is also expressible in percentage.

Consider the decimal fraction .65; this is $\frac{65}{100}$ or 65%. Similarly, $.035 = \frac{35}{1000} = \frac{3.5}{100} = 3.5\% = 3\frac{1}{2}\%$; $.273 = \frac{273}{1000} = \frac{27.3}{100} = 27.3\%$; $.00935 = \frac{935}{100000} = \frac{.935}{100} = .935\%$; etc. That is, .65 = 65%; .035 = 3.5%; .273 = 27.3%; .00935 = .935%; etc. From these examples we see at once the following

RULE: *To express a decimal fraction in percentage, shift the decimal point two places to the right.*

Since the shifting of the decimal point two places is equivalent to multiplying by 100, and since a common fraction is reduced to a decimal by dividing the numerator by the denominator, we have at once the

RULE: *To express a common fraction or ratio in percentage, multiply the numerator by* 100 *(add two zeros) and divide by the denominator.*

The result of any arithmetical calculation involving multiplication or division of fractions or cancellation may be expressed in percentage by thus adding two zeros in the numerator and carrying out the final indicated division.

As an illustration of the second rule, let us express the ratio or fraction $\frac{53}{79}$ as a percentage. We have 5300 ÷ 79 = 67.1%. Similarly, $\frac{163}{725}$ is 16300 ÷ 725 = 22.5% or $22\frac{1}{2}\%$. As another illustration: 375 out of a total of 1845 represents what percentage? It is 37500 ÷ 1845 = 20.3% or $20\frac{3}{10}\%$.

131. Percentage Expressed as a Fractional Part. By reversing the rules given in the last article, any stated percentage may be expressed as a decimal or common fraction of the total on which the percentage is based. Thus, from the first rule we have:

RULE: *To express a percentage as a decimal fraction, shift the decimal point in the percentage number two places to the left.*

As illustrations: 43% = .43; 17.4% = .174; $62\frac{1}{2}\%$ = 62.5% = .625; $3\frac{1}{4}\%$ = 3.25% = .0325; etc.

Since any percentage is a fraction with 100 as denominator it may be expressed as a simple common fraction as follows:

RULE: *To express a percentage as a common fraction, write the percentage number as a numerator with* 100 *as denominator, and reduce the resulting fractions to lowest terms.*

As illustrations: $65\% = \frac{65}{100} = \frac{13}{20}$; $25\% = \frac{25}{100} = \frac{1}{4}$; $62.5\% = \frac{62.5}{100} = \frac{5}{8}$; $3.75\% = \frac{3.75}{100} = \frac{375}{10000} = \frac{3}{80}$; etc.

132. Profit and Loss. Taking *profit* to mean gain in a commercial transaction expressed as a fractional part of the amount of money or value of property involved, and *loss* as similarly expressed, both may very conveniently be stated in percentage. Loss is generally expressed as a percentage of the original amount or value involved, while profit may be expressed as a percentage of either the original or final amount or value.

Thus, suppose a quantity of merchandise is purchased for 100 dollars and sold for 125 dollars. The gain is 25 dollars, and this is $\frac{25}{100}$ = 25% of the original or cost price, or $\frac{25}{125}$ = 20% of the final amount or selling price. This result is stated by saying that the profit *on the investment* is 25 percent, or that 20 percent *of the receipts* is profit.

In a transaction involving buying and selling, this example is an illustration of the following general

RULE: (i) *Subtract cost price from selling price* (*or original from final value*).

(ii) *Multiply remainder by* 100 *and divide by cost* (*or original value*) *to find percentage profit "on investment."*

(iii) *Multiply remainder in* (i) *by* 100 *and divide by selling price* (*or final value*) *to find percentage of profit "in receipts."*

Since loss is always based on cost price (or original value) we have the following

RULE: *Subtract sale price* (*or final value*) *from cost price* (*or original value*), *multiply remainder by* 100, *and divide by cost* (*or original*), *to find percentage loss.*

To illustrate this rule, suppose 125 dollars worth of goods to be sold for 120 dollars. The percentage loss is $100 \times (125 - 120) \div 125 = \frac{500}{125} = 4\%$.

133. Discount. *Discount* is a reduction in a specified price or a reduction from the amount of an acknowledged debt or other obligation. It is always expressed as a percentage but in different cases is expressed on different bases.

If the listed price of an article of sale or merchandise is reduced a certain amount, this amount is stated as a certain percentage of the original or *list* price. Usually the discount in percent is stated alone, and if the actual amount of discount is desired it is to be calculated. Thus, if a certain list price is $75 and the discount is 15% (.15) the actual amount of the discount is .15 × 75 = $11.25. The list price less the discount is called the *net* price. In the example just given, the list price is $75, the discount is 15%, and the net price is $63.75. From this example and discussion we have the

RULE: *To find net price, express the percentage discount as a decimal fraction, multiply the list price by this fraction, and subtract the product from the list price.*

Successive or repeated discounts are sometimes given. This means that after a certain discount has been deducted, the remainder is considered as a new list price and a second discount is deducted from this amount, etc. Thus a discount of "8 and 10" on a list of $325 would mean 325 − (.08 × 325) = 325 − 26 = 299, and 299 − (.10 × 299) = 299 − 29.90 = $269.10, net price. Similarly, a third discount might be stated and calculated as a certain percentage of the $269.10, and so on.

A simpler method of calculation than the one just given can be found as follows: When 8% is deducted 92% of the original remains, and when 10% or .10 of this 92% is deducted this means .10 × 92 = 9.2% of the original, so that there remains 92 − 9.2 = 82.8% or .828 of the original price. The final net price is, therefore, .828 × 325 = $269.10, as before. This illustrates the following rule for successive discounts:

RULE: (i) *Subtract the first percentage discount from* 100 *and multiply the remainder by the second discount expressed as a decimal; then subtract the product from the first remainder.*

(ii) *If there is a third discount proceed as with the second, and so on.*

(iii) *The list price multiplied by the final remainder, expressed as a decimal, is the net price.*

A discount on a debt is an amount deducted for payment before the date on which it is due, or the amount deducted in advance as interest (see next article) agreed upon to be paid, to find the so-called *present worth* of the debt.

Thus, suppose it is agreed that $1000 is to be paid at a certain future time, the $1000 to include a certain percentage as interest on the amount now due; or suppose that a certain amount is borrowed now and when paid with interest at that future time will amount to $1000. In either case, what is the amount *now* due, or the amount borrowed *now?* That is, what is the *present worth?*

This is easily found as follows: Suppose the discount for interest or for advance payment is 6%. This means that the present worth is 100% and the 6% added will amount to $1000 at the specified future date. That is, 106% of the present worth, or 1.06 times the present worth, is $1000. The present worth is therefore $1000 \div 1.06 = 943.396$ dollars, or $943.40.

From this example we have, for the simple case illustrated, the

RULE: *Express the percentage discount as a decimal fraction and add it to* 1.00; *and divide the amount of the debt by this sum. The quotient is the present worth.*

As an illustration: Suppose a man borrows money, agreeing to pay $250 in one year, which is to include interest at 7%. How much cash does he actually receive? According to the rule we have: $7\% = .07$; $1.00 + .07 = 1.07$; $250 \div 1.07 =$ $233.64, cash received.

134. Interest. In business transactions, *interest* is money paid for the use of borrowed money or equivalent value. It is always stated as a percentage of the amount borrowed. This percentage is called the interest *rate;* the amount borrowed is called the *principal;* and the sum of principal and interest is called the *amount.* The amount is therefore the total sum to be paid.

The interest rate is usually expressed as a certain percentage of the principal for *one year* and in the case of *simple* interest the total interest is simply the annual interest multiplied by the number of years the loan continues to "run," that is, the number of years from the time the money is borrowed until the time it is paid.

Suppose $175 is borrowed at 6% interest and is to run $2\frac{1}{2}$ years; what will be the amount due? The amount may be found in several ways; probably the most convenient is the following: In one year 6% is to be added to the principal. Therefore in $2\frac{1}{2}$ years $2\frac{1}{2} \times 6 =$

15% is to be added. This means that 115% of the principal, \$175, or $1.15 \times 175 = \$201.25$ is to be paid. That is, the amount is \$201.25.

If the time to run is stated by giving the number of years and days, or the dates of the loan and payment, the total time is to be reduced (as a compound denominate number) to years and days and the days expressed as decimal fraction of a year by dividing by 365. The total time is then a decimal number of years. From the example and explanations given, therefore, we have at once the following

RULE FOR SIMPLE INTEREST: *Express the time to run as a decimal number of years and multiply by the rate; express the percentage result as a decimal fraction and add to* 1.00; *and multiply the principal by the sum. The result is the amount.*

As an example, let us find the amount of \$750 at 6% simple interest for 3 yr. 115 day. The 115 days are $\frac{115}{365} = \frac{23}{73} = .315$ year and hence the time is 3.315 years. The total percent interest is then $3.315 \times 6 = 19.89\% = .1989$ or .199, and the amount is $750 \times 1.199 = \$899.25$.

In finding the time in interest calculations, a month is usually taken as 30 days.

In simple interest, the interest money for the first year or other interest period does not itself draw interest for the remainder of the time to run. When this is not the case, and the interest does in turn itself draw interest, the interest is said to be *compounded*. The calculation of compound interest is somewhat involved and is discussed separately in the next chapter.

Chapter 20

COMPOUND INTEREST

135. Compound Interest. When a sum of money draws interest for a specified period of time, and at the end of this period the amount of principal and interest is taken as a new principal to draw interest for another period of time, the interest is said to be compounded and is referred to as *compound interest.*

As in simple interest, the original sum is the *principal;* the interest percentage is the *rate;* and the sum of principal and interest is the *amount;* the period or interval of time during which the interest runs before being compounded is the interest *period;* the interest is usually compounded repeatedly at the ends of equal periods throughout the total time to run, or interest *term.* If the interest period is one year, six months, three months, one month, etc., the interest is said to be *compounded annually*, compounded *semi-annually*, *quarterly*, *monthly*, etc., respectively.

The interest rate is based on and stated for the year, as 4% annually. If the interest period is the year it is calculated separately for each year as in simple interest, the principal for each succeeding year being the amount at the end of the preceding year. If the interest is compounded semi-annually (two periods per year) the interest rate *for the period*, called the *periodic rate*, is half the annual rate, and in general the periodic rate is the annual rate divided by the number of periods per year.

136. Calculation of Compound Interest by Simple Percentage. Suppose $500 to draw interest at 6%, compounded annually. What is the amount in two years?

As in simple interest, the interest for one year is $.06 \times 500 = \$30$, and the amount at the end of the first year is therefore $530. Taking this as the principal for the second year, the interest for the second year is $.06 \times 530 = \$31.80$, and the amount at the end of the second year is therefore $530 + 31.80 = \$561.80$, which is the total amount due.

If it is desired to find the amount at compound interest for three or any number of years the calculations just made are repeated for each year, as often as may be necessary.

By using the rule given in article 134 the calculation of compound interest by simple percentage may be somewhat simplified. Thus in one year at 6%, 1 dollar amounts to 1.06 dollars. Taking this $1.06 as principal for the second year it will at the end of the second year amount to $1.06 \times 1.06 = 1.1236$, and $500 will amount to $500 \times 1.1236 =$ $561.80, as before. Similarly, in three years the amount of 1 dollar would be $1.06 \times 1.06 \times 1.06 = 1.191$ and the amount of $500 would be $500 \times 1.191 =$ $595.50.

Examining these results, it is noticed that the amount of 1 dollar at 6% in *one* year is 1.06; in *two* years it is $1.06 \times 1.06 = (1.06)^2$; in *three* years it is $1.06 \times 1.06 \times 1.06 = (1.06)^3$; similarly in *four* years it would be $(1.06)^4$; etc. The entire amount is then the principal multiplied by the amount for one dollar. If the rate is 4%, 5%, or any other, we would have 1.04, $(1.04)^2$, $(1.04)^3$, etc. These results and the general method of calculation by simple percentage can be concisely stated somewhat as follows

RULE I. *To find the amount of any sum compounded annually at any rate: express the rate as a decimal fraction and add to* 1.00; *raise the sum to a power whose exponent is the number of years; and multiply the result by the principal. The product is the amount.*

If the compounding is not annual, the calculation is based on the number of *periods* of any length and the periodic rate is used. This gives in the same way

RULE II. *To find the amount of a sum at any compound interest: divide the annual rate by the number of periods per year and express the periodic rate so obtained as a decimal fraction; add this fraction to* 1.00 *and raise the sum to a power whose exponent is the total number of interest periods; and multiply the result by the principal. The product is the amount.*

When the compounding is annual, Rule II becomes the same as Rule I, but both are stated for convenience.

As an example of the use of Rule II, let us calculate the amount of $300 compounded quarterly at 8% per year for $1\frac{1}{2}$ years. Here the periodic rate is $8 \div 4 = 2\% = .02$, and the total number of periods is $1\frac{1}{2} \times 4 = 6$. The amount of 1 dollar is therefore $(1.02)^6 = 1.126$ and the total amount is $300 \times 1.126 =$ $337.80.

137. Calculation of Compound Interest by Geometrical Progression. The rules of the preceding article show that the amount of any principal is found by multiplying the principal by 1 plus the periodic rate, as many times as there are interest periods. That is, the principal is multiplied by the multiplier (1 plus periodic rate), this product is multiplied by the same multiplier, this product in turn by the same multiplier, and so on.

Now this is exactly the way the successive terms in a geometrical progression are formed (article 69). Therefore, a question in compound interest amounts to a question in *geometrical progression*, in which the

First term is the *principal;*
Common ratio is 1 + *periodic rate;*
Number of terms is the *number* of *periods* + 1;
Last term is the *amount.*

Any compound interest problem can, therefore, be solved by the rules of article 70.

By using the rules of article 70, therefore, and changing the names of the elements into the corresponding names used in compound interest, not only can the amount be calculated but when any three of the four quantities *principal*, *rate*, number of *periods*, or *amount* are specified the fourth can be found directly.

We shall not here give examples of the various compound interest calculations as these have been illustrated in the geometrical progression examples.

138. Calculation of Compound Interest by Logarithms. The compound interest rules developed in article 136 really amount to the rule for calculating the last term in a geometrical progression, and as seen in article 70, this calculation is very conveniently carried out by means of logarithms. Thus, to find the amount of \$300 at 8% compounded quarterly for $1\frac{1}{2}$ years (the example worked out in article 136) we have: *Amount* = $300 \times (1.02)^6$, and to raise 1.02 to the 6th power is a simple logarithmic operation, even though very laborious by straightforward multiplication.

To carry out the calculation logarithmically, we have:

$$\begin{aligned} \log 1.02 &= 0.0086 \\ 6 \times \log 1.02 &= 0.0516 \\ (1.02)^6 = \text{anti-log} &= 1.126 \\ \text{Amount} &= 1.126 \times 300 \\ &= \$337.80 \end{aligned}$$

In this the number, 1.02 is 1+ periodic rate, and the periodic rate is the annual rate divided by the number of periods per year; the power 6 to which the 1.02 is raised is $1\frac{1}{2} \times 4$, the time in years multiplied by the number of periods per year; and the multiplier 300 is the principal. The multiplication by the principal could also be performed logarithmically simply by adding the logarithm of 300 to 6 times the logarithm of 1.02, and looking up the final antilogarithm, which would be the amount.

The calculation of compound interest for any rate, term and manner of compounding can therefore be carried out logarithmically by the following

RULE: *Express the periodic rate as a decimal and add to* 1.00; *multiply the logarithm of the sum by the number of interest periods, and to the product add the logarithm of the principal; the antilogarithm of the result is the desired amount.*

As an example of this method, let us find the amount of \$1275 at 6% compounded semi-annually for 5 years.

Here the periodic rate is $6 \div 2 = 3\% = .03$, the number of periods is $5 \times 2 = 10$ and the principal is 1275. Adding $1.00 + .03 = 1.03$, therefore, the calculation is as follows:

$$\begin{aligned} \log 1.03 &= 0.0128 \\ 10 \times \log 1.03 &= 0.1280 \\ +\log 1275 &= 3.1055 \\ \hline \log \text{Amt.} &= 3.2335 \\ \text{Amt.} &= \$1712. \end{aligned}$$

By applying logarithms to the rules of article 70, any of the problems mentioned in article 137 can be solved by means of logarithms.

139. Compound Interest Table. The three methods explained in articles 136, 137, and 138 for calculating compound interest are all based on finding the amount of one dollar and multiplying this amount by the principal. If, therefore, the amount of one dollar were known for any periodic rate and any number of periods, the calculation of compound interest would be simple. The amounts of one dollar have been calculated by logarithms by the method of the preceding article for different rates and periods and the results are given in the *compound interest table.* With this table available, it is, therefore, only necessary to read off the amount of one dollar for a specified rate and number of periods and multiply this amount by the principal, and the problem is solved.

The dollar amounts are calculated by raising the number 1+

periodic rate to the power indicated by the number of periods, by means of logarithms and the rule of article 53. These amounts are listed in Table V at the end of this book, on the line with the number of periods and in the column under the periodic rate. Thus, the amount of $1 in 20 periods at the periodic rate of $3\frac{1}{2}\%$ is $1.98979. In this manner the one dollar amount can be found for any rate and term within the range of the table.

140. How to Use the Compound Interest Table. The description of the compound interest table given in the preceding article shows how to find the amount of any sum at any rate (up to 7% per period) compounded in any manner for any term (up to 50 periods). For the sake of completeness, however, the method of use of the table is here concisely stated as

RULE I. *From the term, rate and manner of compounding, determine the periodic rate and number of periods; in the table on the line with the periods and in the column under the periodic rate, find the amount of one dollar; and multiply this amount by the principal. The product is the desired total amount.*

Since the total amount is the dollar amount multiplied by the principal, the principal required to produce any specified amount is that amount divided by the dollar amount. We have, therefore,

RULE II. *To find the principal required to produce a specified amount at compound interest, determine the periods and periodic rate as in* RULE I *and from the table find the amount of one dollar; the specified amount divided by the dollar amount is the required principal.*

If it is required to find the time in which a certain sum will produce a stated amount at compound interest, consideration of the last two rules and the description of the table show that it may be found as follows:

RULE III. *Divide the stated amount by the principal to determine the amount of one dollar, and also determine the periodic rate; in the table under this rate locate the dollar amount just determined; on the same line and at the left is the required number of periods.*

If it is required to find the rate at which a certain sum will produce a stated amount in a given time, use

RULE IV. *Determine the amount of one dollar as in* RULE III, *and also the number of periods; on the line with the periods locate this dollar amount; at the head of the column is the required periodic rate.*

We will give here an example illustrating the use of each of the four rules just given.

Example I. Suppose the annual rate is 8% and the compounding is quarterly; find the amount of \$100 in 4 years.

The periodic rate is $8 \div 4 = 2\%$ and the number of periods is $4 \times 4 = 16$. In the table under 2% and on the line with 16 periods is the dollar amount \$1.37279. The amount of \$100 is then $100 \times 1.37279 = \$137.28$.

Example II. What principal is required to produce an amount of \$1000 in 10 years at 6% compounded semi-annually?

The number of periods is $10 \times 2 = 20$ and the periodic rate is $6 \div 2 = 3\%$. From the table the amount of one dollar in 20 periods at 3% is \$1.80611. The principal is, therefore, by Rule II, $1000 \div 1.806 = \$553.71$.

Example III. In what time will \$100 double itself at 8% compounded quarterly?

Here the principal is \$100 and the amount \$200; the dollar amount is therefore \$2, and the periodic rate is $8 \div 4 = 2\%$. From the table, the dollar amount at 2% is \$1.99989, or \$2, in 35 periods, and since these are quarterly the time is $8\frac{3}{4}$ years, or 8 yr. 9 mo.

Example IV. Find the rate at which \$250 will produce \$1350 in 43 years, compounded annually.

The amount of one dollar in the 43 annual periods is $1350 \div 250 = \$5.40$, and in the table 5.40 is found on the line with 43 periods under the periodic (here annual) rate of 4%.

This completes the solutions of all the usual problems in compound interest.

141. Remarks. In commercial and actuarial mathematics the principles and methods of percentage and interest are studied very completely and very large and complete tables are used. This and the preceding chapter are intended to give only an introduction to the subject, and it is believed that the explanations and the sample problems and calculations are sufficient to enable any one to solve any problem he may meet in ordinary work. For that reason no exercises are included for solution.

Chapter 21

BINARY ARITHMETIC

142. Counting by Tens—The Denary System. The basis of ordinary arithmetic is the denary system. The word denary is derived from the Latin word for ten, and implies that ten characters are used. These characters are 1, 2, 3, 4, 5, 6, 7, 8, 9 and 0. By means of these ten characters alone, without using any other symbol except the decimal point, any rational real number, however large or small, may be expressed.

The method by which this is done has become so familiar by daily use that it is taken for granted. Yet it was unknown to many earlier civilizations, and its invention was one of the great milestones in the history of mathematics.

To approach this method, write the first nine integral numbers with their corresponding symbols:

Number	one	two	three	four	five	six	seven	eight	nine
Symbol	1	2	3	4	5	6	7	8	9

then consider how the next integral number, ten, is written. It is done by starting over again with the symbol 1, and moving it one place to the left, using the empty (or null) symbol 0 to fill the vacant place, and so to show that the 1 has been moved. Thus the number ten is represented by the symbols 1 and 0 grouped in second and first positions, as shown:

Second position ↘ ↙ First position

10

The higher numbers than ten are now written successively by replacing the 0 successively by 1, 2, 3, 4, 5, 6, 7, 8, and 9.

Number	*Symbol*	*Number*	*Symbol*	*Number*	*Symbol*
eleven	11	fourteen	14	seventeen	17
twelve	12	fifteen	15	eighteen	18
thirteen	13	sixteen	16	nineteen	19

From this point the process of expressing the successively higher numbers follows a strictly logical process. Each group of ten numbers from this point, that is, number twenty through twenty-nine, thirty through thirty-nine, and so on, are expressed by writing the symbols 2, 3, and so on, in the second position, and then changing the symbols in the first position successively from 0 to 1, 1 to 2, until 9 is reached just as was done in writing the numbers from ten to nineteen.

The number ninety-nine clearly is the greatest number that can be expressed by this two-position system. Therefore the next greater integral number is expressed by extending the positional system to a third position, writing there the symbol 1, and writing the null symbol 0, both in the second and first positions, to give 100.

Then integral numbers successively greater than one hundred are written by repeating the sequence of steps already taken for the numbers up to 100, and extending the operation, when 199 is reached, to the third position by writing 200. When 999 is reached, a fourth position is required; when 9999 is reached, a fifth position is necessary, and so on. Thus the decimal system expresses any integral number, however great, in terms of ten symbols. By introducing the decimal point, and using positions to the right of it, the denary system is extended to express any rational number, however small, and is then called by the better-known name, decimal system.

143. Counting by Eights—The Octal System. The possibilities of this invention of positional writing of numbers are not confined to the denary system. Suppose that the symbols 8 and 9 did not exist. Any number great, could still be written using only the symbols, 1, 2, 3, 4, 5, 6, 7, and 0.

Such a system would simply use the second position for the number one unit greater than 7, that is, eight would be written 10, and the first twenty-four integral numbers in such a system would be written

one	1	nine	11	seventeen	21
two	2	ten	12	eighteen	22
three	3	eleven	13	nineteen	23
four	4	twelve	14	twenty	24
five	5	thirteen	15	twenty-one	25
six	6	fourteen	16	twenty-two	26
seven	7	fifteen	17	twenty-three	27
eight	10	sixteen	20	twenty-four	30

By extending the table, it is found that the symbol 100 corresponds to the number sixty-four. This can be inferred directly from the fact that this system has only eight symbols, instead of ten for the decimal systems. Thus the octal system has, in its multiplication table up to and including the symbol 100, eight rows of eight numbers each (instead of ten rows of ten numbers each) and therefore 100 in the octal system represents eight times eight, or sixty-four (instead of ten times ten, or one hundred).

144. Counting by Fours. The method by which the system of counting by eights was inferred from the decimal system can readily be extended to systems with any other numbers of symbols. The method involved, the use of an empty or null symbol, zero, with symbols for integers starting with 1, the symbol for one, and extending to a symbol for the number one less than the total numbers of symbols. The method involved the use of the single symbols for the integral numbers up to the largest number for which the particular system has a symbol; the expression of the next higher number by the symbol for unity, 1, followed by the null symbol, 0, the formation of the successively higher numbers by replacing the 0 with the single symbols in order continuing the process to three positional numbers, four positional numbers, . . . , as necessary.

Thus in counting by fours—the symbols used are 0, 1, 2 and 3. The numbers one, two and three are written 1, 2 and 3; four is written 10; five, 11; six, 12; seven, 13; eight, 20; nine, 21; ten, 22; eleven, 23; twelve 30; sixteen 100, seventeen 101, and so on. Note that the symbol-group 100 is reached for the number four times four in the four-system, just as it is reached for the number eight times eight in the eight-system, and for the number ten times ten in the denary system.

145. Counting by Twos or the Binary System. The method of establishing counting systems is very readily used to form a two-system. The only symbols required are 0 and 1. The number one is written 1; the number two, 10; the number three, 11; and the number 4, 100. Note that four is two times two, so that this symbol-group 100 occurs at the same relative point as in the other systems. Greater numbers are given in the more complete table on page 218.

This binary-system table has been given in greater length than those for the other systems, and the remainder of this chapter is devoted to it, because of the many important applications of the binary system. They arise wherever "yes-or-no" situations occur. Since these are

BINARY SYSTEM TABLE

Number	*Symbol or Symbol-group*	*Number*	*Symbol or Symbol-group*
One	1	Sixteen	10000
Two	10	Thirty-Two	100000
Three	11	Sixty-Four	1000000
Four	100	One Hundred Twenty-Eight	10000000
Five	101	Two Hundred Fifty-Six	100000000
Six	110	Five Hundred Twelve	1000000000
Seven	111	One Thousand and Twenty-Four	10000000000
Eight	1000	Two Thousand and Forty-Eight	100000000000

inherent in logical propositions, which are concerned with reasoning about truth or falsity, binary arithmetic (and its counterpart Boolean algebra, which is treated in "Algebra for the Practical Man") is basic to mathematical logic. It is also well adapted to digital computers, which are so widely used today, and to other devices having integrated switching systems.

Since binary arithmetic uses only two symbols, 0 and 1, then binary numbers can be represented directly by two-state systems, such as simple switches, which have only two positions, "open" or "closed"; or by electron tubes or transistors functioning as switches. One such switch can be used to represent each place or position in the binary number. This advantage more than compensates for the larger number of positions that are needed to express a number in the binary system than in the ordinary denary one.

Moreover, the additional number of positions needed for large numbers is not as great as might be inferred from the number of positions needed for small numbers. For example, the number two is written 10 in the binary system and the number four is written 100 so that the second position accommodates only two numbers, two and three. On the other hand, the number one thousand and twenty-four is written 10000000000 and the number two thousand and forty-eight is written 100000000000, so that the eleventh position accommodates one thousand and twenty-four numbers. This property of the binary system is due to the property of positional notation whereby the symbol-groups 10, 100, 1000, and so on represent the first, second, third, etc.

powers of the base number. Thus, in the binary system 10 symbolizes two; 100 symbolizes two × two (2^2) or four; 1000 symbolizes two × two × two (2^3) or eight; . . . 10000000000 symbolizes 2^{10} or one thousand and twenty-four, 100000000000 symbolizes 2^{11}, or two thousand and forty-eight.

146. Converting Denary Numbers into Binary Numbers. Up to this point in the discussion of the binary system numbers have been expressed in binary symbols only, by writing the number itself in words. This has been done to avoid confusion with the denary (ordinary) system notation. From here on, however, both will be expressed in digits, with, of course, a clear specification of the system in every instance.

In order to convert an integral number from any number system to another, one could use a table such as that on page 218 for the denary to binary system. However, this trouble can be avoided by following this simple rule.

RULE 1. *Divide repeatedly the integral number to be converted by the base of the new system* (2 *for the binary system*) *and copy the remainders in reverse order.*

As an example, convert the denary number 29 to the binary system.

2	29	
2	14	Remainder 1
2	7	Remainder 0
2	3	Remainder 1
2	1	Remainder 1
	0	Remainder 1

The number is written 11101 in the binary system.

This result can be checked by use of the table on page 218 by taking the values of the symbol-groups in the binary system.

Binary	Decimal
10000	16
1000	8
100	4
1	1
11101	29

As a second example, convert the denary number 57 to the binary system.

2	57	
2	28	Remainder 1
2	14	Remainder 0
2	7	Remainder 0
2	3	Remainder 1
2	1	Remainder 1
	0	Remainder 1

The number is written 111001.

Note that this method is limited by Rule 1 to integral (whole) numbers to be converted. If the number is a fraction, the procedure is different. The first step is to obtain the fraction in proper decimal form, if it is not so already, by the methods explained earlier in this book. Then apply the following rule.

RULE 2. *Multiply repeatedly the denary decimal fraction to be converted by the base of the new system (2 for the binary system), subtract the whole numbers from the product (in the binary system, 1 is the only whole number that can appear), continue until a zero result is obtained, or if not, until the desired number of decimal places have been obtained, and copy the subtracted whole numbers in true order.*

As example, convert the denary number $\frac{13}{16} = .8125$ into the binary system.

.8125	
2	
1.625	
1.	Subtract 1
.625	
2	
1.25	
1.	Subtract 1
.25	
2	

```
  .50
          No subtraction (write 0)
 ____
  .50
    2
 ____
 1.
 1.       Subtract 1
 ____
 0
```

The number is written as .1101.

Note that repeated multiplication of the denary number .8125 by 2 and subtraction of 1, finally produced a zero result, showing that the corresponding binary number .1101 was exactly equal to the denary number .8125. However, for most denary decimal fractions, no number of multiplications by 2 and subtractions of 1 will yield a zero result. That is, most denary numbers cannot be expressed exactly in the binary system. In that case, the conversion is carried out to as many places as the accuracy of the work requires. In this book, they will be carried out to eight significant figures, in order to facilitate checking by the student of his results to the exercises (at the end of the chapter) with the answers (at the end of the book).

Two points are to be emphasized, however, in regard to this incomplete conversion. The binary number is always less than the denary number that was converted, since there is always an unconverted residue. While by continuing the conversion to a great enough number of places the difference can be made extremely small, it never vanishes.

The other point is that the number of binary significant figures necessary for a given degree of precision is not uniform, depending, among other considerations, upon the number of significant figures in the denary decimal number. The use of eight significant figures, while convenient for the purposes of this book, is therefore not a working rule. As an example, convert the denary number .43 into the binary system.

```
  .43
    2
 ____
  .86
          No subtraction (write 0)
 ____
  .86
    2
 ____
```

1.72
1. Subtract 1
.72
2
1.44
1. Subtract 1
.44
2
.88
No subtraction (write 0)
.88
2
1.76
1. Subtract 1
.76
2
1.52
1. Subtract 1
.52
2
1.04
1. Subtract 1
.04

The binary number is written as .0110111 . . . The conversion process has been discontinued at this point at only six significant figures, because the next two figures will be zero.

As another example, convert the denary decimal number .031 into the binary system.

.031
2
.062
No subtraction (write 0)

```
 .062
    2
 ----
 .124
        No subtraction (write 0)
 ----
 .124
    2
 ----
 .248
        No subtraction (write 0)
 ----
 .248
    2
 ----
 .496
        No subtraction (write 0)
 ----
 .496
    2
 ----
 .992
        No subtraction (write 0)
 ----
 .992
    2
 -----
1.984
1.      Subtract 1
 -----
 .984
    2
 -----
1.968
1.      Subtract 1
 -----
 .968
    2
 -----
1.936
1.      Subtract 1
 -----
 .936
    2
 -----
1.872
1.      Subtract 1
 -----
```

.872	
2	
1.744	
1.	Subtract 1
.744	
2	
1.488	
1.	Subtract 1
.488	
2	
.976	
	No subtraction (write 0)
.976	
2	
1.952	
1.	Subtract 1
.952	

The binary number is written as .0000011111101 . . .

The next, and final, type of denary-to-binary conversion to be considered is that if a mixed denary number. The method is expressed in the following rule.

RULE 3. *Separate the mixed number into its integral and (decimal) fractional parts. Convert the integral part by Rule* 1, *the decimal part by Rule* 2, *and add the two results.*

As an example, convert the denary number 19.71 into the binary system.

2	19	(Integral part of denary number)
2	9	Remainder 1
2	4	Remainder 1
2	2	Remainder 0
2	1	Remainder 0
	0	Remainder 1

The integral part of the denary number is 10011 in the binary system.

.71	(Fractional part of mixed number)
2	
1.42	
1.	Subtract 1
.42	
2	
.84	
	No subtraction (write 0)
.84	
2	
1.68	
1.	Subtract 1
.68	
2	
1.36	
1.	Subtract 1
.36	
2	
.72	
	No subtraction (write 0)
.72	
2	
1.44	
1.	Subtract 1
.44	
2	
.88	
	No subtraction (write 0)
.88	
2	
1.76	
1.	Subtract 1
.76	

The fractional part of the denary number is .10110101 . . . in the binary system.

Thus by Rule 3, the denary number 19.71 is given in the binary system by

10011	Integral part
.10110101 . . .	Fractional part
10011.10110101 . . .	Complete binary number (to eight places after the binary point)

147. Converting Binary Numbers Into Denary Numbers. The direct method of converting binary numbers into denary numbers consists merely of writing, for each position in the binary number, its denary equivalent. In the table given earlier in this chapter, it was shown that the symbol 1 in the binary system was equivalent to 1 in the denary system; 10 in the binary system to 2 in the denary system; binary 100, to denary 8; binary 1000 to denary 16, and so on, each 0 added to the binary number increasing the denary equivalent value by a factor of two. Therefore, the rule for direct conversion of binary integral numbers to denary numbers is the following.

RULE 4. *Starting with the right-hand side of the binary number, write* 1 *if there is a* 1 *in that position,* (*nothing of course, if there is a* 0). *Then write* 2 *if there is a* 1 *in the second binary position; write* 4 *for a* 1 *in the third binary position, and continue this process for the successive binary positions* (*from the right*) *by writing* 8, 16, 32, *and successively higher powers of two for each binary position in which a* 1 *appears. Add these figures to obtain the required denary number.*

As an example, convert the binary number 10101101 to the corresponding denary number.

For 1 in the right-hand position, write	1
For 0 in the second from right-hand position, no entry is made	
For 1 in the third from right-hand position, write	4
For 1 in the fourth from right-hand position, write	8
For 0 in the fifth from right-hand position, no entry is made	
For 1 in the sixth from right-hand position, write	32
For 0 in the seventh from right-hand position, no entry is made	
For 1 in the eighth from right-hand position, write	128
Add the figures written	173

The 173 is the denary equivalent of the binary number 10101101.

The method of converting binary fractional numbers to denary numbers is analogous to the preceding method, for converting binary integral numbers. The difference in the case of fractions is that negative powers of 2, instead of positive powers are used, as expressed in the following rule.

RULE 5. *Starting with the position just to the right of the binary point, write* $\frac{1}{2} = 0.5$ *if there is a* 1 *in that position (nothing, of course, if there is a* 0*). Then write* $\frac{1}{4} = 0.25$ *if there is a* 1 *in the second position to the right of the binary point; write* $\frac{1}{8} = 0.125$ *for the third position, and continue the process for the successive positions by writing* $\frac{1}{16} = 0.0625$, $\frac{1}{32} = 0.03125$, $\frac{1}{64} = 0.015625$ *and successively lower powers of* 2 *for each position to the right in which a* 1 *appears. Add these figures to obtain the required denary number, and even off the result to a number of places consistent with the desired or known precision.*

As an example, convert the binary number .10101101 to the corresponding denary number.

For 1 in the first position after the binary point, write $\frac{1}{2}$ =	.5
For 0 in the second position after the binary point, no entry is made	
For 1 in the third position after the binary point, write $\frac{1}{8}$ =	.125
For 0 in the fourth position after the binary point, no entry is made	
For 1 in the fifth position after the binary point, write $\frac{1}{32}$ =	.03125
For 1 in the sixth position after the binary point, write $\frac{1}{64}$ =	.015625
For 0 in the seventh position after the binary point, no entry is made	
For 1 in the eighth position after the binary point, write $\frac{1}{256}$ =	.00390625
Add the figures written	.67578125

As stated in the rule, the binary number .10101101 thus corresponds to the denary number .67578125, which is then shortened to the number of places consistent with the known or desired precision.

As in the case of conversion of denary to binary mixed numbers, the conversion of binary mixed numbers to denary numbers requires four operations, as stated in the following rule.

RULE 6. *Separate the mixed binary number into its integral and fractional parts. Convert the integral part by Rule* 4; *the fractional part by Rule* 5, *and add the two results.*

As an example, convert the binary number 11001.1101 into the denary system.

11001 (Integral part of binary number).

For 1 in the right-hand position, write	1.
For 0 in the second from right position, no entry is made.	
For 0 in the third from right position, no entry is made.	
For 1 in the fourth from right position, write	8.
For 1 in the fifth from right position, write	16.
Add the figures written.	25

Thus 25 is the denary equivalent of the binary number 11001.

.1101 (Fractional part of binary number)

For 1 in the first position after the binary point, write $\frac{1}{2}$ =	.5.
For 1 in the second position, write $\frac{1}{4}$ =	.25.
For 0 in the third position, no entry is made.	
For 1 in the fourth position, write $\frac{1}{16}$ =	.0625.
Add the figures written.	.8125

Thus .8125 is the denary equivalent of the binary number .1101. Then by adding 25 and .8125, there is obtained the denary number 25.8125, which is the equivalent of the binary number 11001.1101.

148. Addition of Binary Numbers. In adding binary numbers, the first step is to write the numbers in column, as in adding numbers in any other system. If some or all of the numbers to be added are fractional or mixed numbers, then they must be written in column so that the binary points are aligned vertically. After the binary numbers have been written in column, they are added, position by position, starting at the right hand side.

In adding the binary numbers in a column, follow the rule for the binary system, that is, $0 + 0 = 0$; $0 + 1 = 1$; $1 + 1 = 10$ (write the zero and carry the 1; $1 + 1 + 1 = 11$ (write the 1 and carry the 1); $1 + 1 + 1 + 1 = 100$ (write 0 and carry the 10), and so on. The method of binary addition is expressed in the following rule.

RULE 7. *To add binary numbers, write them with binary points aligned vertically (or, in the case of integral binary numbers, with the right-hand digits aligned vertically). Then add the columns, starting at the right, writing below each column the right-hand digit* (0 *or* 1) *of the partial total and carrying the other(s) to the next column or columns.*

As an example, add the binary numbers 011 and 1001.

$$\begin{array}{r} 011 \\ 1001 \\ \hline 1100 \end{array}$$

Thus, the addition of the right-hand column is $1 + 1 = 10$, write the 0 and carry the 1; the next column is 1 (carry) $+ 1 = 10$, write the 0 and carry the 1; the next column is 1 (carry) $+ 0 + 0 = 1$, write the 1, and the final (left) column is simply the 1.

As another example, add the binary numbers 11, 10, 110, 100, and 1100.

$$\begin{array}{r} 11 \\ 10 \\ 110 \\ 100 \\ 1100 \\ \hline 11011 \end{array}$$

Thus, the addition of the right-hand column gives simply 1, write the 1; the next column is $1 + 1 + 1 = 11$, write the 1 and carry the 1; the next column is 1 (carry) $+ 1 + 1 + 1 = 100$, write the 0 and carry the 10; the next column is 10 (carry) $+ 1 = 11$, write the 1 and carry the 1; the last entry is 1 (carry) with no digits above, write the 1.

As still another example, add the binary numbers 11.01, 110.11, 1111.111.

$$\begin{array}{r} 11.01 \\ 110.11 \\ 1111.111 \\ \hline 11001.111 \end{array}$$

The first right-hand column has only one digit, 1, write the 1; the next column has $1 + 1 + 1 = 11$, write the 1 and carry the 1; the third column has 1 (carry) $+ 1 + 1 = 11$, write the 1 and carry the 1;

the fourth column has 1 (carry) + 1 + 1 = 11, write the 1 and carry the 1; the fifth column has 1 (carry) + 1 + 1 + 1 = 100, write the 0 and carry the 10; the sixth column has 10 (carry) + 1 + 1 = 100, write the 0 and carry the 10, the seventh column has 10 (carry) + 1 = 11, write the 1 and carry the 1; the eighth position has 1 (carry), write the 1.

149. Subtraction of Binary Numbers. Subtraction of binary numbers is done in the same way as in the denary system. The subtrahend (the number to be subtracted) is written below the minuend (the number to be subtracted from) with binary points directly one below the other (or in case of integral numbers, with the right-hand digits aligned). Then subtraction is done position by position, starting from the right. Where a 1 occurs below a 0, a 1 is "borrowed" from the next position to the left in the upper number; if there is a 0 in this position then a 1 in the second position to the left is transferred, as in "borrowing" of denary arithmetic. Subtraction is summarized in the following rule.

RULE 8. *For subtraction, align the binary numbers vertically, and subtract position by position, starting from the right. If a* 0 *occurs above a* 1, *"borrow" from the successive positions to the left of the minuend as necessary.*

As an example, subtract 10111 from 100111.

$$\begin{array}{r} 100111 \\ 10111 \\ \hline 10000 \end{array}$$

The sequence of steps, starting from the right side, are 1 − 1 = 0, 1 − 1 = 0, 1 − 1 = 0, 0 − 0 = 0, and 10 − 1 = 1.

As another example, subtract 1 from 100.

$$\begin{array}{r} 100 \\ 1 \\ \hline 11 \end{array}$$

Here the sequence of steps are 100 = 10 + 10 = 10 + 1 + 1. Then 10 + 1 + 1 − 1 = 10 + 1 = 11.

As a third example, subtract 111.0101 from 1010.111.

$$\begin{array}{r} 1010.111 \\ 111.0101 \\ \hline 11.1001 \end{array}$$

Here the sequence of steps are, starting from the right side, $10 - 1 = 1$, $0 - 0 = 0$, $1 - 1 = 0$, $1 - 0 = 1$, and to the left of the binary point, $10 - 1 = 1$, leaving 1000 as the remainder of the minuend after using the 10.111, so that $1000 = 100 + 100 = 100 + 10 + 10$ remaining in the minuend; while the partial subtrahend, after the 1. was used, becomes 110. or $100 + 10$. Then $100 + 10 + 10 - 100 + 10 = 10$, so that the difference is 11.1001.

150. Multiplication of Binary Numbers. Multiplication of binary numbers follows the same method as that of denary numbers. Moreover, since the only symbols are 1 and 0, the operation is very simple, the only precaution necessary being to keep the partial products in the correct position. Because of the large number of zeros, they can get out of position somewhat easily. One safeguard is to write a full line of zeros for each zero in the multiplier.

The rule is simply stated.

RULE 9. *For multiplication, write multiplicand and multiplier as in denary arithmetic and follow the same process of writing partial products and adding them. The rule for multiplication is simply* $1 \times 1 = 1$, $1 \times 0 = 0$.

As an example, multiply 1101 by 1011.

```
    1101
    1011
    ----
    1101
   1101
  0000
 1101
--------
10001111
```

As another example, multiply 10111 by 10001.

```
    10111
    10001
    -----
    10111
   00000
  00000
 00000
10111
---------
110000111
```

As still another example, multiply 111.11 by 100.101.

```
        111.11
        100.101
        -------
        111 11
       0000 0
      11111
     00000
    00000
   11111
   ------------
   100011.11011
```

Note that since the multiplicand has two significant fiugres after the binary point, and the multiplier, three, then five places are pointed off in the product.

151. Division of Binary Numbers. Like their multiplication, the division of binary numbers follows the same procedure as the division of denary numbers. The division is subtracted repeatedly from the dividend until there is no remainder, or until the required or desired number of significant figures have been obtained. If there is still a remainder after the last of the digits of the dividend have been used, a zero is added to the remainder for the next subtraction of the divisor, and operation is continued until there is no remainder or until as above, the required or desired number of significant figures have been obtained.

RULE 10. *For division, write division and dividend as in denary arithmetic and follow the same process of repeated subtraction of the divisor, bringing down another digit from the dividend after each subtraction to form a new minuend, continuing with zeros after the last digit in the dividend has been used.*

As an example, divide 10101 by 111.

```
       11
111)10101
    111
    ---
     111
     111
     ---
```

As another example, divide 100110 by 1101.

```
         110
  111)1001110
      1101
       1101
       1101
```

As a third example, divide 11100.111 by .1011.

```
              101010.
   .1011)11100.1110
         1011
          1101
          1011
            1011
            1011
```

Note that a zero was added to the dividend above to bring up its number of places after the binary point to the number in the divisor, and so to establish the position of the decimal point in the quotient.

152. Exercises and Problems.

1. Convert to the binary system the denary number 15.
2. Convert to the binary system the denary numbers 25; 35; 45.
3. Convert to the binary system the denary decimal number .25.
4. Convert to the binary system the denary decimal numbers .55; .74; 85.
5. Convert to the binary system the denary decimal numbers .40; .11; .062.
6. Convert to the binary system the denary fraction $\frac{1}{2}$.
7. Convert to the binary system the denary fractions $\frac{1}{4}$; $\frac{1}{8}$; $\frac{1}{3}$.
8. Convert to the binary system the denary mixed number 12.5.
9. Convert to the binary system the denary mixed numbers 9.7; 13.31; 19.246.
10. Convert to the binary system the denary mixed numbers 10.1; 20.2; 30.3.
11. Convert to the denary system the binary number 101.
12. Convert to the denary system the binary numbers 111; 1111; 11111.
13. Convert to the denary system the binary numbers 101; 1001; 10001; 100001.
14. Convert to the denary system the binary number 10.11.
15. Convert to the denary system the binary numbers 111.1; 111.11; 111.111.

16. Convert to denary system the binary numbers 101.1; 100.101; 10101.10101.
17. Add the binary numbers 10 and 110.
18. Add the binary numbers 1101, 1011, and 10011.
19. Add the binary numbers 1, 101, 1001 and 10001.
20. Add the binary numbers 11.1 and 11.11.
21. Add the binary numbers 10.1; 101.101 and 11001.01011.
22. Add the binary numbers 100.11, 1000.11 and 10001.11.
23. Convert the denary numbers 7, 8 and 9 into binary numbers, add these, and convert the result to the denary number.
24. Subtract 1010 from 110101.
25. Subtract 1111 from 10000.
26. Subtract 101.101101 from 1000.
27. Subtract 1111.00001 from 10101.1.
28. Multiply 1001 by 1100.
29. Multiply 11111 by 10111.
30. Multiply 11001 by 1111.11.
31. Divide 101111001 by 1101.
32. Divide 1001101.1 by 11.111.
33. Divide 1011110.1 by 100.11.

REVIEW PROBLEMS

1. Find the distance from home plate to second base if the standard baseball field is a square of side 90 feet. (Answer correct to the nearest hundredth.)
2. Eric took 12 minutes to walk $\frac{5}{8}$ of a mile. At this rate how many minutes will it take him to walk the rest of the mile?
3. A 36-centimeter board is cut into two pieces which are in the ratio of 4:5. Find the length of each piece.
4. A mathematician offers her newly hired assistant a choice of two methods of payment for a job which will take 12 hours.

   Method I: a flat rate of $1,000 for the whole job.

   Method II: 25¢ for the first hour, 50¢ for the second hour, $1 for the third hour and so on, doubling the amount made each hour.

   Which payment method will result in the higher salary?
5. If 3 pounds of apples cost 89¢, how much will $2\frac{1}{4}$ pounds of apples cost?
6. A motorist travels at 55 miles per hour for 3 hours to get to Boston. How long would it have taken him if he had traveled at 45 miles per hour?
7. An economically minded driver finds that his car gets 20 miles per gallon when he travels at a speed of 50 miles per hour but only 18 miles per gallon when he travels at 60 miles per hour. If gasoline costs $1.50 per gallon, find the cost of making a 360-mile trip at each speed.
8. Find the volume and the total surface area of a circular cylinder with radius 3 centimeters and height 7 centimeters. (Answer may be left in terms of π.)
9. a. Which has the greater volume, a cube of edge 5 inches or a sphere of radius 3 inches?

   b. Which has the greater surface area?
10. If the edge of a cube is doubled, what happens to its surface area? What happens to its volume? (Compare the results obtained using cubes of edges 3 and 6.)
11. A chemist has 100 grams of an element whose half-life is 15 days.

    a. How many grams of the element remain after 60 days?

    b. In how many days will there be less than 10 grams remaining?
12. A new jogger is able to run for 12 minutes the first day. He wishes to increase his jogging time by 2 minutes per day. How many minutes will he jog after 3 weeks?
13. How much would it cost to install wall-to-wall carpeting at $18 per square yard in a room whose dimensions are 12 feet by 17 feet 6 inches?
14. Find the circumference of a circle that is inscribed in a square of perimeter 24. (Leave your answer in terms of π.)
15. A rectangular pool measures 30 feet by 75 feet. The law requires that a concrete walkway 4 feet wide completely surround the pool. If concrete costs $1.10 per square foot installed, how much would it cost to install the walk?

16. Find the area of a circle which circumscribes a square of perimeter 24. (Leave answer in terms of π.)
17. A rectangular sheet of tin measures 6 inches by 21 inches. Metal is to be cut from this sheet to construct a closed cylindrical can of radius 2 inches and height 5 inches. Correct to the nearest square inch, how many square inches of tin are wasted? (Use $\pi = 3.14$.)
18. A van uses 18.2 gallons of gasoline to make a trip of 310 miles. Calculate the mileage per gallon correct to the nearest tenth.
19. The annual interest rate on a loan is 18%. If the loan is repaid in less than a year, the interest rate is prorated. If 5,000 is borrowed for three months, compute the amount of interest charged.
20. A person has a sum of $10,000 to place in a bank account for a period of one year. Bank *A* offers $10\frac{1}{3}\%$ simple interest. Bank *B* offers 10% interest compounded quarterly. Which is the wiser investment and by how much?
21. A person's monthly bank statement for her checking account shows an opening balance of $423.86. She made deposits of $280, $310, $260, and $300. She wrote checks in the amounts of $38.48, $512.40, $54.23, $62.50, $182.69, $94.75, $12.08, and $214.81. What is her closing balance?
22. Express each of the following base 10 numbers in base 2:

    a. 17 b. 32 c. 44

    Express each of the following base 2 numbers in base 10:

    d. 11001 e. 1110 f. 100111000
23. Perform each of the indicated operations in the binary system:

    a. 101101 + 100011
    b. 101101 − 10011
    c. 1010 × 111
24. A man invested $8000 in two separate companies. He makes a profit of 4% on the first investment and takes a loss of 3% on the second. If he makes a net profit of $110, how much did he invest in each company?
25. A company pays each worker $5.20 per hour up to 40 hours per week and pays "time and a half" for every hour of overtime. Calculate the total week's pay for each of the following workers:

    a. Mr. Smith who worked 40 hours
    b. Ms. Ray who worked 46 hours
    c. Mr. Mann who worked 36 hours
26. A piece worker earns whichever of the following two quantities is greater: $1.20 per piece or $4 per hour. Each person works an 8-hour day. Calculate the minimum number of finished pieces a worker must produce in a day so that his salary exceeds that paid on an hourly basis.
27. Find the G. C. D. and the L. C. M. of the following pairs of numbers:

    a. 84 and 275 c. 612 and 312
    b. 15 and 90 d. 253 and 154
28. Express each of the following decimals as a fraction in lowest terms:

    a. 0.36 c. 0.3666666. . .
    b. 0.333333. . . d. 0.36363636. . .

29. Perform the indicated operations and leave the results in simplest form:
    a. $3\frac{2}{3} + 6\frac{3}{5} + 11\frac{1}{2}$
    b. $12\frac{3}{8} - 7\frac{5}{6}$
    c. $2\frac{1}{4} \times 1\frac{7}{12} \times 2\frac{2}{3}$
    d. $1\frac{7}{9} \div \frac{8}{15}$
30. Perform the indicated operations:
    a. $32.1 + 5.438 + 18.72 - 20.84$
    b. 115.06×5.7
    c. $9.3 \div 2.03$
31. Express each of the following as a percent:
    a. $\frac{3}{4}$
    b. $\frac{5}{8}$
    c. $\frac{3}{2}$
    d. $\frac{2}{3}$
    e. 0.62
    f. 0.2
    g. 0.05
    h. 0.003
    i. 3.15
32. If a certain city has an 8% sales tax, calculate the total cost of an item that is marked:
    a. 69¢
    b. $3.29
    c. $29.95
33. If another city has a 5% sales tax and a shopper has exactly $10.00 in his wallet, what is the highest price that he can pay for an item and not run out of money?
34. Express as a decimal correct to the nearest hundredth:
    a. $\frac{8}{11}$
    b. $1\frac{3}{8}$
    c. $\frac{13}{9}$
    d. $\frac{2}{5}$
35. The cost of mailing a letter by first-class mail is 15¢ for the first ounce and 11¢ for each additional ounce. How much would it cost to mail a 7-ounce letter?
36. A sheet of plywood is 4 feet by 8 feet. A woman wishes to construct an open box with a base of 3 feet by 3 feet and sides of 3 feet by 1 foot. How many square feet of plywood will be left over?
37. Which of the following numbers has the greatest value?
    a. $3\frac{1}{8}$
    b. 3.14
    c. $\frac{22}{7}$
    d. 315%
38. A sales item retails for $100. Which would be a better bargain, a single discount of 24% or a discount of 15% followed by a discount of 10%? Why?
39. A kitchen floor is 100 inches by 150 inches. Solid-color tiles come in boxes of 12 where each tile is 1-foot square. What is the minimum number of boxes a person must buy to cover the floor?
40. If you have 30 feet of fencing available, which would enclose the greater area, a square plot or a circular plot?
41. Which of the following quantities is larger, $30 \times 42 \times 26$ or $22 \times 63 \times 20$? (Do not multiply out.)
42. Which is the better buy for the consumer in the supermarket, a 14-ounce bottle of ketchup at 89¢ or a 40-ounce bottle of the same brand of ketchup for $2.79?
43. Express each of the following Fahrenheit temperatures in Celsius:
    a. 131
    b. 14
    c. 48
    d. 100
44. Express each of the following Celsius temperatures in Fahrenheit:
    a. 35
    b. 80
    c. –15
    d. 68

45. Convert each of the following amounts of time to seconds:
    a. 8 hours
    b. 1 hour 30 minutes
    c. 4 hours, 28 minutes, 15 seconds
46. There are 4.75 French francs to the American dollar.
    a. You exchange a $20 traveller's check. How many francs do you get?
    b. An item in a Parisian department store sells for 120 francs. Find the equivalent cost in dollars.
47. In one gas station, unleaded gas costs $1.40 per gallon while in another gas station the same gas costs 40¢ per liter. If you need 10 gallons of gasoline, which station is cheaper and by how much?
48. The legal speed limit is 55 miles per hour. What would the legal limit be if it were expressed in kilometers per hour?
49. In a given week the daily sale of stocks is illustrated by the graph.
    a. In millions, find the average number of shares sold per day.
    b. On which day did the number of shares sold show the greatest increase over the previous day?

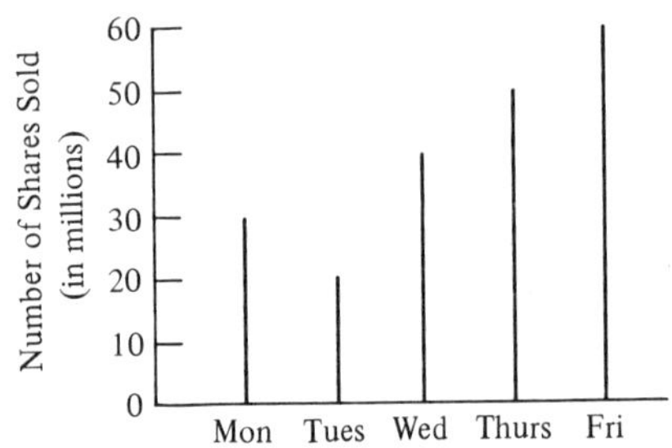

50. The graph below represents the budget of a small city. From the data shown calculate the percentage of the budget allocated to education.

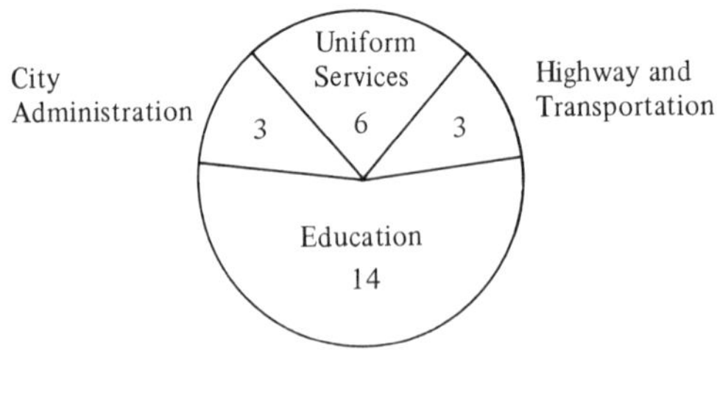

ANSWERS TO PROBLEMS

1. 127.28 ft.
2. 7.2 min.
3. 16 cm., 20 cm.
4. Method II will yield the higher salary by \$23.75.
5. 67¢.
6. 3 hrs. 40 min.
7. \$27 @ 50 mph, \$30 @ 60 mph.
8. 63 π, 60 π.
9. a. the cube.
b. the cube.
10. The area is multiplied by 4, the volume is multiplied by 8.
11. a. 6.25 gms.
b. 51 days.
12. 52 min.
13. \$420.
14. 6 π.
15. \$994.40.
16. 18 π.
17. 38 sq. in.
18. 17.0 mi. per gal.
19. \$225.
20. Bank *B*'s investment is wiser by \$4.80.
21. \$401.92.
22. a. 10001.
b. 100000.
c. 101100.
d. 25.
e. 14.
f. 312.
23. a. 1010000.
b. 11010.
c. 1000110.
24. He invested \$5,000 @ 4% and \$3,000 @ 3%.
25. a. \$208.
b. \$254.80.
c. \$187.20.
26. 27 pieces.
27.

	G. C. D.	L. C. M.
a.	1	23,100.
b.	15	90.
c.	12	15,912.
d.	11	3,542.

28. a. $\frac{25}{9}$.
b. $\frac{1}{3}$.
c. $\frac{11}{30}$.
d. $\frac{4}{11}$.
29. a. $\frac{653}{30}$.
b. $\frac{109}{24}$.
c. $\frac{19}{2}$.
d. $\frac{10}{3}$.
30. a. 35.418.
b. 655.842.
c. 4.58.
31. a. 75%.
b. 62.5%.
c. 150%.
d. $66\frac{2}{3}$%.
e. 62%.
f. 20%.
g. 5%.
h. .3%.
i. 315%.
32. a. 75¢.
b. \$3.56.
c. \$32.35.
33. \$9.52.

34. a. .73.
b. 1.38.
c. 1.44.
d. .40.

35. 81¢.

36. 11 sq. ft.

37. d.

38. The single discount of 24% is better by 50¢.

39. 9 boxes.

40. the circular plot.

41. 30 × 42 × 26.

42. The 14-oz. bottle is the better buy.

43. a. 55.
b. −10.
c. 9.
d. 38.

44. a. 95.
b. 176.
c. 5.
d. 154.

45. a. 28,800.
b. 5,400.
c. 16,095.

46. a. 95 francs.
b. $25.26.

47. gallon station by $1.14.

48. 88.5.

49. a. 38.
b. Wednesday.

50. 54%.

ANSWERS TO EXERCISES

Article 19, Page 26

1. 1676.77.
2. 6.0.
3. 805.2.
4. 31.83.
5. 50.99 in.
6. 8.4 in.
7. $87,646.84.
8. 12.4 m.
9. 1283.07 in.
10. 93.20 sq. in.
11. 82.0 cu. in.
12. 5.47 in.

Article 26, Page 39

4. 13.
5. 28.
6. 7.
7. 798.
8. 6300.
9. 360.
10. 11 ft.
11. $3528.
12. 17 mi. 5115 ft.
14. 6930.
15. 150, 100, 75, 60, 50.
16. $12.60.
17. 28 men each, 77 squads.
18. 12 in.

Article 35, Page 52

1. $\frac{3}{5}$; $\frac{7}{8}$; $\frac{2}{3}$; $\frac{5}{9}$; $\frac{3}{4}$.
2. $17\frac{1}{9}$; $26\frac{11}{13}$; $24\frac{1}{2}$; $17\frac{13}{28}$; $4\frac{121}{209}$.
3. $\frac{79}{5}$; $\frac{127}{8}$; $\frac{99}{4}$; $\frac{143}{7}$; $\frac{101}{12}$.
4. $\frac{18}{24}$, $\frac{20}{24}$ and $\frac{21}{24}$; $\frac{24}{30}$, $\frac{25}{30}$, $\frac{27}{30}$; $\frac{72}{168}$, $\frac{70}{168}$, $\frac{52}{168}$.
5. $2\frac{1}{4}$.
6. $3\frac{1}{8}$.
7. $1161\frac{49}{180}$.
8. $\frac{1}{4}$.
9. $\frac{11}{24}$.
10. $\frac{1}{15}$.
11. $24\frac{11}{18}$.
12. $70\frac{4}{7}$.
13. $2\frac{2}{3}$; $\frac{2}{5}$; 6.
14. $\frac{10}{21}$; $\frac{63}{80}$; $\frac{3}{5}$.
15. $\frac{7}{24}$.
16. $\frac{1}{15}$.
17. 2.
18. $\frac{3}{35}$; $\frac{2}{11}$; $\frac{5}{17}$.
19. 35; 12; 20.
20. $1\frac{1}{12}$; $1\frac{13}{15}$; $\frac{3}{4}$.
21. 36; $1\frac{2}{3}$; $1\frac{5}{7}$.
22. 4.
23. 80.
24. 32.
25. .5833+; .9375; .48; .828125.
26. $\frac{17}{20}$; $\frac{3}{8}$; $\frac{127}{400}$; $\frac{108}{125}$; $\frac{11}{16}$.
27. $\frac{1}{7}$; $\frac{2}{3}$; $\frac{571}{625}$+.

Article 43, Page 64

1. 737,881; 132,651; 35,831,808.
2. 194,481; 148,035,889; 43,046,721.
3. 319; 1234; 26.54; 63.34.
4. 98; 123; 18.9; 15.89.
5. 75; 27; 7.

Article 57, Page 83

1. 1008; 1027; 860.6; 3.942; .1660; 5,332,000.
2. 26.76; 2.16; 225.1; .5216; .5746; 17,400.
3. 2304; 246.5; 38,416; 861,200; 269,100; 57,800.

4. 96; 212.5; 6; 3; 35.07; 17.37.
5. 26.96; 156.0; .3815; 3.522.

Article 65, Page 91

1. $19\frac{1}{2}$.
2. 9.
3. 18.
4. 245.
5. 16.
6. \$184.24.
7. 23.
8. $113\frac{1}{7}$ ft.
9. 100 ft.
10. 550 lb.
11. 20 in.
12. 403 ft.
13. 500 ft.
14. 48.2 gal.
15. 5.

Article 72, Page 102

1. 33.
2. $7\frac{19}{24}$.
3. 2.
4. 15.
5. 154.
6. 1st payment, \$50; amount, \$1485.
7. 6144.
8. 4.
9. 7.
10. 765.
11. 280.
12. 986.5.
13. 7.
14. 1st day, 20 mi.; total, 650 mi.
15. 10 days.
16. 1st, \$20; last, \$164.
17. Debt, \$10,930; 7 payments.
18. \$5000.
19. 300.
20. 156.
21. 43 ft.
22. 30 in.
23. 26th week, \$335,544.32; total, \$671,088.63.
24. 331.968 ft.
25. 128 times as many as at start.
26. $\frac{1}{9}$; $\frac{1}{3}$; $\frac{7}{33}$; $\frac{647}{990}$; $\frac{89}{99}$.

Article 86, Page 124

1. 6,336,000 in.
2. 1600 rd.
3. \$183.60.
4. 517.5 mi.
5. 5760 sq. rd.
6. \$9,169.60.
7. 157.
8. \$82.50.
9. 2 ft. $8\frac{2}{5}$ in.
10. 262 mi. 6 fur. 16 rd.
11. 3 mi. 1 fur. 17 rd. 2 yd. 1 ft. 8 in.
12. 70 cd. 2 cd. ft. 8 cu. ft.
13. 27,072.
14. \$2640.75.
15. 5.504 mi.
16. 2 T. 3 cwt. 97 lb. 6 oz.
17. 16 yd. 2 ft. 1 in.
18. 62 bu. 1 pk. 5 qt. $1\frac{2}{3}$ pt.
19. 7 lb. 10 oz. 14 dwt. 6 gr.
20. 11 rd. $11\frac{3}{4}$ in.
21. 33 mi. 2 fur. 33 rd. $7\frac{1}{2}$ ft.
22. 2 bu. 1 pk. 4 qt.
23. 125,000 g.; 17,000,000 mm.; 5730 cm.^{2}; 27,000 dm.^{3}.
24. 55.125 lb.; 186.45 mi.; 68.705 qt.; 18.928 l.; 80.465 Km.
25. 984.3 ft.; .1864 mi.
26. 380.9 m.; .237 mi.
27. 44 Kg.; 97 lb.
28. $83\frac{1}{3}$ cm.; 32.8 in.
29. $16\frac{2}{3}$.
30. 3.01 sec.; 4.85 sec.

Article 93, Page 135

1. 9,486,720 min.
3. 325,060″.
4. 4 dy. 23 hr. 50 min. 5 sec.
5. 300 dy.
6. 1 yr. 274 dy. 4 hr. 30 min.
7. 7 yr. 9 mo. 1 dy.
8. 5 yr. 161 dy. 9 hr. 22 min.
9. 10° C.

10. 10,832° F.
12. 327° C.
13. 39.2° F.
14. 3600″; 1,296,000″.
15. 21,600′.
16. 55,695″.
17. 60° 12′ 26″.
18. $\frac{1}{8}$.
19. 24°; 1440′.
20. (a) 30°, 360°.
(b) 30′, 6°.
21. (a) 6°, 360°.
(b) 6′, 6°.

Article 101, Page 151

1. 1592 mi.
2. 5259 mi.
3. 2756 mi.
4. 7095 mi.
5. 90° 15′ W.
6. 71° 3′ W.
7. 82° 22′ W.
8. 85° 5′ W.
9. 29 min. 32 sec.
10. 6 hr. 37 min. 40 sec.
11. 7 hr. 15 min. $11\frac{13}{15}$ sec.
12. 11 hr. 1 min.
13. 3604 mi.
14. 4647 mi.
15. 833 mi.
16. 5 hr. 34 min.; 4188 mi.
17. 14,166 mi., approx.

Article 110, Page 166

1. 50.625 A.
2. 11 ft.
3. $1029.00.
4. 210 sq. ft.
5. 600 sq. ft.
6. 13.261 A.
7. 349.08 sq. ft.
8. 20 in.
9. 39 ft.
10. 25.61 ft.
11. 28.28 ft.
12. 7.71 ft.
13. 5.236 ft.
14. 15.9 ft.
15. 5.89 ft.
16. $24\frac{1}{3}$ ft.
17. 672.
18. 78.54 sq. ft.
19. 113.1 sq. in.
20. 114.59 A.
21. 8.73 sq. ft.
22. 392.7 sq. ft.
23. $282.75.
24. 10 ft.; 16.42 ft.; 866 sq. ft.
25. 2.64 mi.

Article 117, Page 174

1. 60 cu. ft.; 106 sq. ft.
2. 12,600 lb.
3. 48.214 bu.
4. (i) 12.9 A.; (ii) 104,167 sq. yd.; (iii) 3,472,222 cu. yd.
5. 11 ft. 7 in.
6. 576 ft.
7. $200.

Article 121, Page 181

1. 11.48 sq. ft. = 11 sq. ft. 69 sq. in.
2. 263.89 sq. ft.
3. 73.63 cu. ft.
4. 824.67 sq. ft.
5. 512.9 sq. ft.
6. 39.27 cu. ft.
7. 254.47 sq. in.
8. 314.16 sq. ft.
9. D = 7929.8 mi.; S = 197,545,000 sq. mi.;
V = 261,078,800,000 cu. mi.
10. 5875.2 gal.
11. 53 in.
12. 9.38 in.
13. 27.
14. 53.86 bu.
15. 49; 6.85 sq. ft.
16. 47.6 cu. in.
17. 3.8112 cu. in.
18. 231.8 cu. in.
19. 1197 gal.
20. 5.603 cu. in.

Article 127, Page 201

1. 8.65 in.
2. 197 sq. in.
4. 1958–1959.
5. 1959–1960.
6. 62,500.
8. 1.875 hr. = 1 hr. 52 min. 30 sec.
9. 20 mi.
10. 1 and 5 hr. after second train starts.

Article 152, Page 233

1. 1111.
2. 11001; 100011; 101101.
3. .01.
4. .100011 . . . ; .11; .11011001 . . . .
5. .0110011 . . . ; .00011100001 . . . ; .000011111101 . . . .
6. .1.
7. .01; .001; .01010101 . . . .
8. 1100.1.
9. 101.10110011 . . . ; 1101.01001111 . . . ; 10011.0011111011 . . . .
10. 1010.000110011 . . . ; 10100.00110011 . . . ; 11110.010011001 . . . .
11. 5.
12. 7; 15; 31.
13. 5; 9; 17; 33.
14. 2.75.
15. 7.5; 7.75; 7.875.
16. 5.5; 4.625; 21.65625.
17. 1000.
18. 101011.
19. 100000.
20. 111.01.
21. 1000.001.
22. 11111.01.
23. The binary numbers are 111, 1000 and 1001; their sum is 11000; the denary equivalent is 24.
24. 101011.
25. 1.
26. 10.010011.
27. 101.01111.
28. 1101100.
29. 1011001001.
30. 110001001.11.
31. 11101.
32. 10100.
33. 10011.111001 . . . .

TABLES

TABLE I

Multiplication and Division

The product of any number in the first row and any number in the first column is found at the intersection of the row and column containing the two numbers.

1	2	3	4	5	6	7	8	9	10	11	12	13	14	15	16	17	18	19	20
2	4	6	8	10	12	14	16	18	20	22	24	26	28	30	32	34	36	38	40
3	6	9	12	15	18	21	24	27	30	33	36	39	42	45	48	51	54	57	60
4	8	12	16	20	24	28	32	36	40	44	48	52	56	60	64	68	72	76	80
5	10	15	20	25	30	35	40	45	50	55	60	65	70	75	80	85	90	95	100
6	12	18	24	30	36	42	48	54	60	66	72	78	84	90	96	102	108	114	120
7	14	21	28	35	42	49	56	63	70	77	84	91	98	105	112	119	126	133	140
8	16	24	32	40	48	56	64	72	80	88	96	104	112	120	128	136	144	152	160
9	18	27	36	45	54	63	72	81	90	99	108	117	126	135	144	153	162	171	180
10	20	30	40	50	60	70	80	90	100	110	120	130	140	150	160	170	180	190	200
11	22	33	44	55	66	77	88	99	110	121	132	143	154	165	176	187	198	209	220
12	24	36	48	60	72	84	96	108	120	132	144	156	168	180	192	204	216	228	240
13	26	39	52	65	78	91	104	117	130	143	156	169	182	195	208	221	234	247	260
14	28	42	56	70	84	98	112	126	140	154	168	182	196	210	224	238	252	266	280
15	30	45	60	75	90	105	120	135	150	165	180	195	210	225	240	255	270	285	300
16	32	48	64	80	96	112	128	144	160	176	192	208	224	240	256	272	288	304	320
17	34	51	68	85	102	119	136	153	170	187	204	221	238	255	272	289	306	323	340
18	36	54	72	90	108	126	144	162	180	198	216	234	252	270	288	306	324	342	360
19	38	57	76	95	114	133	152	171	190	209	228	247	266	285	304	323	342	361	380
20	40	60	80	100	120	140	160	180	200	220	240	260	280	300	320	340	360	380	400

TABLE II

Primes,* Multiples and Factors (Part 1)

N	01	07	11	13	17	19	23	29	31	37	41	43	47	49	53	59	61	67	71	73	77	79	83	89	91	97
000	..	..	..	..	..	..	..	..	..	..	..	..	..	7	..	..	..	..	..	..	7	..	..	..	7	..
300	7	..	..	..	..	11	17	7	..	..	11	7	..	..	..	..	19	..	7	..	13	..	..	..	17	..
600	..	..	13	..	..	..	7	17	..	7	..	..	..	11	..	..	..	23	11	..	..	7	..	13	..	17
900	17	..	..	11	7	..	13	..	7	..	..	23	..	13	..	7	31	..	..	7	..	11	..	23	..	..
1200	..	17	7	..	..	23	..	..	..	..	17	11	29	..	7	..	13	7	31	19	..	..	..	..	..	..
1500	19	11	..	17	37	7	..	11	..	29	23	..	7	..	..	..	7	..	..	11	19	..	..	7	37	..
1800	..	13	..	7	23	17	..	31	..	11	7	19	..	43	17	11	..	..	..	..	..	..	7	..	31	7
2100	11	7	..	..	29	13	11	..	..	..	..	..	19	7	..	17	..	11	13	41	7	..	37	11	7	13
2400	7	29	..	19	..	41	..	7	11	..	..	7	..	31	11	..	23	..	7		..	37	13	19	47	11
2700	37	..	..	..	11	..	7	..	..	7	..	13	41	..	..	31	11	..	17	47	..	7	11	..	..	..
3000	..	31	..	23	7		..	13	7	..	..	17	11	..	43	7	..	..	37	7	17	..	..	..	11	19
3300	..	..	7	..	31	..	.	..	..	47	13	..	..	17	7	..	..	7	..	..	11	31	17	..	..	43
3600	13	..	23	..	..	7	..	19	..	..	11	..	7	41	13	..	7	19	..	..	..	13	29	7	..	..
3900	47	..	..	7	..	..	..	..	..	31	7	..	..	11	59	37	17	..	11	29	41	23	7	..	13	7
4200	..	7	..	11	..	..	41	..	..	19	..	..	31	7	..	..	..	17	..	..	7	11	..	..	7	..
4500	7	..	13	..	..	..	..	7	23	13	19	7	..	..	29	47	..	..	7	17	23	19	..	13	..	..
4800	..	11	17	..	..	61	7	11	..	7	47	29	37	13	23	43	..	31	..	11	..	7	19	..	67	59
5100	..	..	19	..	7	..	47	23	7	11	53	37	..	19	..	7	13	..	..	7	31	..	71	..	29	..
5400	11	..	7	..	..	..	11	61	..	..	..	..	13	..	7	53	43	7	..	13	..	..	..	11	17	23
5700	..	13	..	29	..	7	59	17	11	..	..	..	7	..	11	13	7	73	29	23	53	..	..	7	.	11
6000	17	.	..	7	11	13	19	..	37	..	7	..	..	23	..	73	11	..	13	..	59	..	7	..	..	7
6300	..	7	..	59	..	71	..	..	13	..	17	..	11	7	..	..	..	..	23	..	7	..	13	..	7	..
6600	7	..	11	17	13	..	37	7	19	..	29	7	17	61	..	..	..	59	7	..	11	..	41	..	..	37
6900	67	..	.	31	..	11	7	13	29	7	11	53	..	..	17	..	..	..	..	19	..	7	..	29	..	..
7200	19	..	..	..	7	..	31	..	7	..	13	..	..	11	..	7	53	13	11	7	19	29	..	37	23	..
7500	13	..	7	11	..	73	..	..	17	..	..	19	..	..	7	..	..	7	67	..	..	11	..	..	..	71
7800	29	37	73	13	..	7	..	..	41	17	..	11	7	47	..	29	7	..	17	..	..	..	..	7	13	53
8100	..	11	..	7	..	23	..	11	47	79	7	17	..	29	31	41	..	..	..	11	13	..	7	19	..	7
8400	31	7	13	47	19	..	..	..	..	11	23	..	..	7	79	11	..	..	43	37	7	61	17	13	7	29
8700	7	..	31	..	23	..	11	7	..	..	..	7	..	13	..	19	..	11	7	31	67	..	..	11	59	19
9000	..	..	..	..	71	29	7	..	11	7	..	..	83	..	11	..	13	..	47	43	29	7	31	61	..	11
9300	71	41	..	67	7	..	..	19	7	..	..	..	13	..	47	7	11	17	..	7	..	83	11	41	..	..

*The *smallest* prime factors of multiple numbers are given. Prime numbers are indicated by dots.

Example: To find the prime factors of 2413: The smallest prime factor, from above table, is 19; then, $2413 \div 19 = 127$. Now, from Part 2, on the following page, 127 is found to be a prime number. Hence, the prime factors of 2413 are 19 and 127.

TABLE II

Primes,* Multiples and Factors (Part 2)

N	01	03	07	09	13	19	21	27	31	33	37	39	43	49	51	57	61	63	67	69	73	79	81	87	91	93	97	99
100	..	..	..	..	..	7	11	..	..	7	..	..	11	..	..	..	7	..	..	13	..	..	..	11	..	..	..	..
400	..	13	11	..	7	..	..	7	..	..	19	..	..	..	11	..	..	..	..	7	11	..	13	..	..	17	7	..
700	..	19	7	..	23	..	7	..	17	..	11	..	..	7	..	..	..	7	13	..	..	19	11	..	7	13	..	17
1000	7	17	19	..	..	..	..	13	..	..	17	..	7	..	.	7	..	..	11	..	29	13	23	..	..	..	..	7
1300	..	..	..	7	13	..	..	..	11	31	7	13	17	19	7	23	..	29	..	37	..	7	..	19	13	7	11	..
1600	..	7	..	..	..	..	..	..	7	23	..	11	31	17	13	..	11	..	..	..	7	23	41	7	19	..	..	..
1900	..	11	..	23	..	19	17	41	..	..	13	7	29	..	..	19	37	13	7	11	..	..	7	..	11	..	..	..
2200	31	..	..	47	..	7	..	17	23	7	..	..	..	13	..	37	7	31	..	..	..	43	..	..	29	..	..	11
2500	41	..	23	13	13	11	..	7	..	17	43	..	..	..	..	..	13	11	17	7	31	..	29	13	..	..	7	23
2800	..	..	7	53	29	..	7	11	19	..	..	17	..	7	..	..	..	7	47	19	13	..	43	..	7	11	..	13
3100	7	29	13	..	11	..	..	53	31	13	..	43	7	47	23	7	29	..	..	..	19	11	..	..	..	31	23	7
3400	19	41	..	7	..	13	11	23	47	..	7	19	11	..	7	..	..	..	..	..	23	7	59	11	..	7	13	..
3700	..	7	11	..	47	..	61	..	7	..	37	..	19	23	11	13	..	53	..	..	7	..	19	7	17	..	..	29
4000	..	..	..	19	..	..	..	..	29	37	11	7	13	..	..	..	31	17	7	13	..	..	7	61	..	..	17	..
4300	11	13	59	31	19	7	29	..	61	7	..	..	43	..	19	..	7	..	11	17	..	29	13	41	..	23	..	53
4600	43	..	17	11	7	31	..	7	11	41	..	..	..	..	..	..	59	..	13	7	..	..	31	43	..	13	7	37
4900	13	..	7	..	17	..	7	13	..	..	..	11	..	7	..	..	11	7	..	..	..	13	17	..	7	..	19	..
5200	7	11	41	..	13	17	23	..	..	..	..	13	7	29	59	7	..	19	23	11	..	..	..	17	11	67	..	7
5500	..	..	..	7	37	..	..	..	..	11	7	29	23	31	7	..	67	..	19	..	..	7	..	37	..	7	29	11
5800	..	7	..	37	..	11	..	..	7	19	13	..	..	..	..	..	..	11	..	..	7	..	..	7	43	71	..	17
6100	..	17	31	41	..	29	..	11	..	..	17	7	..	11	..	47	61	..	7	31	..	37	7	23	41	11	..	..
6400	37	19	43	13	11	7	..	..	59	7	41	47	17	..	..	11	7	23	29	.	..	11	..	13	..	43	73	67
6700	..	..	19	..	7	..	11	7	53	..	..	23	11	17	43	29	..	..	67	7	13	..	..	11	..	..	7	13
7000	..	47	7	43	.	..	7	..	79	13	31	..	..	7	11	..	23	7	37	..	11	..	73	19	7	41	47	31
7300	7	67	..	..	71	13	..	17	..	..	11	41	7	.		7	17	37	53	..	73	47	11	83	19	..	13	7
7600	11	..	..	7	23	19	..	29	13	17	7	..	..	..	7	13	47	79	11	..	..	7	..	..	..	7	43	..
7900	..	7	..	11	41	..	89	..	7	..	..	17	13	..	..	73	19	..	31	13	7	79	23	7	61	..	11	19
8200	59	13	29	..	43	..	..	19	..	..	..	7	..	73	37	23	11	..	7	..	.	17	7	..	..	..	..	43
8500	..	11	47	67	..	7	..	..	19	7	..	..	..	83	17	43	7	.	13	11	..	23	.	31	11	13	..	..
8800	13	..	..	23	7	..	..	7	..	11	..	..	37	..	53	17	..	..	..	7	19	13	83	..	17	..	7	11
9100	19	..	7	..	13	11	7	..	23	..	..	13	41	7	..	..	..	7	89	53	.	67	..	..	7	29	17	..
9400	7	..	23	97	..	..	..	11	..	..	..	..	7	11	13	7	..	..	..	17	..	..	19	53	..	11	..	7

*The *smallest* prime factors of multiple numbers are given. Prime numbers are indicated by dots.

Example: To find the prime factors of 1001: The smallest prime factor, from above table, is 7; then, $1001 \div 7 = 143$, the smallest prime factor of which is 11; then, $143 \div 11 = 13$, a prime number Hence, the prime factors of 1001 are 7, 11 and 13.

TABLE II

PRIMES,* MULTIPLES AND FACTORS (PART 3)—*Concluded*

N	03	09	11	17	21	23	27	29	33	39	41	47	51	53	57	59	63	69	71	77	81	83	87	89	93	99
200	7	11	..	7	13	..	..	..	..	..	..	13	..	11	..	7	..	..	..	..	..	..	7	17	..	13
500	..	..	7	11	..	..	17	23	13	7	..	..	19	7	..	13	..	..	..	..	7	11	..	19	..	..
800	11	..	..	19	..	..	..	..	7	..	29	7	23	..	..	..	..	11	13	..	..	..	..	7	19	29
1100	..	..	11	..	19	..	7	..	11	17	7	31	..	..	13	19	..	7	..	11	..	7	..	29	..	11
1400	23	..	17	13	7	..	..	..	..	..	11	..	..	..	31	..	7	13	..	7	..	..	..	..	..	..
1700	13	..	29	17	..	..	11	7	..	37	..	..	17	..	7	..	41	29	7	..	13	..	..	..	11	7
2000	..	7	..	..	43	7	..	..	19	..	13	23	7	..	11	29	..	..	19	31	..	..	..	..	7	..
2300	7	..	..	7	11	23	13	17	..	..	..	..	..	13	..	7	17	23	..	..	..	..	7	..	..	..
2600	19	..	7	..	..	43	37	11	..	7	19	..	11	7	..	..	..	17	..	..	7	..	..	..	..	..
2900	..	..	41	..	23	37	..	29	7	..	17	7	13	..	..	11	..	..	..	13	11	19	29	7	41	..
3200	..	..	13	..	..	11	7	..	53	41	7	17	..	..	..	..	13	7	..	29	17	7	19	11	37	..
3500	31	11	..	..	7	13	..	..	..	..	..	..	53	11	..	..	7	43	..	7	..	..	17	37	..	59
3800	..	13	37	11	..	..	43	7	..	11	23	..	..	..	7	17	..	53	7	..	..	11	13	..	17	7
4100	11	7	..	23	13	7	..	..	..	..	41	11	7	..	..	..	23	11	43	..	37	47	53	59	7	13
4400	7	..	11	7	..	..	19	43	11	23	..	..	..	61	..	7	..	41	17	11	..	..	7	67	..	11
4700	..	17	7	53	..	..	29	..	..	7	11	47	..	7	67	..	11	19	13	17	7	..	..	..	..	..
5000	..	..	..	29	..	..	11	47	7	..	71	7	..	31	13	..	61	37	11	..	..	13	..	7	11	..
5300	..	..	47	13	17	..	7	73	..	19	7	..	..	53	11	23	31	7	41	19	..	7	..	17	..	..
5600	13	71	31	41	7	..	17	13	43	..	..	..	..	..	..	..	7	..	53	7	13	..	11	..	..	41
5900	..	19	23	61	31	..	..	7	17	..	13	19	11	..	7	59	67	47	7	43	..	31	..	53	13	7
6200	..	7	..	..	..	7	13	..	23	17	79	..	7	13	..	11	..	..	..	..	11	61	..	19	7	..
6500	7	23	17	7	..	11	61	..	47	13	31	..	..	..	79	7	..	..	..	..	..	29	7	11	19	..
6800	..	11	7	17	19	..	..	..	..	7	..	41	13	7	..	19	..	..	..	13	7	..	71	83	61	..
7100	..	..	13	11	..	17	..	..	7	11	37	7	..	23	17	..	13	67	71	..	43	11	..	7	..	23
7400	11	31	..	..	41	13	7	17	..	43	7	11	..	29	..	..	17	7	31	..	..	7	..	..	59	..
7700	..	13	11	..	7	..	..	59	11	71	..	61	23	..	..	..	7	17	19	7	31	43	13	..	..	11
8000	53	..	..	..	13	71	23	7	29	..	11	13	83	..	7	..	11	..	7	41	..	59	..	..	..	7
8300	19	7	..	..	53	7	11	..	13	31	19	17	7	..	61	13	..	..	11	..	17	83	..	..	7	37
8600	7	..	79	7	37	..	..	..	89	53	..	..	41	17	11	7	..	..	13	..	..	19	7	..	..	..
8900	29	59	7	37	11	..	79	..	..	7	..	23	..	7	13	17	..	..	..	47	7	13	11	89	17	..
9200	..	..	61	13	..	23	..	11	7	..	..	7	11	19	..	47	59	13	73	..	..	..	37	7	..	17
9500	13	37	..	31	..	89	7	13	..	..	7	..	..	41	19	11	73	7	17	61	11	7	..	43	53	29

*The *smallest* prime factors of multiple numbers are given. Prime numbers are indicated by dots.

Example: To find the prime factors of 3211: The smallest prime factor, from above table, is 13; then, 3211÷13=247, the smallest prime factor of which is 13; then, 247÷13=19, a prime number. Hence, the prime factors of 3211 are 13, 13 and 19.

TABLE III

Squares, Cubes, Square Roots, Cube Roots, of Numbers

No.	Square	Cube.	Sq. Rt.	Cu. Rt.	No.	Square	Cube.	Sq. Rt.	Cu. Rt.
0	0	0	0.0000000	0.0000000	65	42 25	274 625	8.0622577	4.0207256
1	1	1	1.0000000	1.0000000	6	43 56	287 496	.1240384	.0412401
2	4	8	.4142136	.2599210	7	44 89	300 763	.1853528	.0615480
3	9	27	.7320508	.4422496	8	46 24	314 432	.2462113	.0816551
4	16	64	2.0000000	5874011	9	47 61	328 509	.3066239	.1015661
5	25	125	2.2360680	1.7099759	70	49 00	343 000	8.3666003	4.1212853
6	36	216	.4494897	.8171206	1	50 41	357 911	.4261498	.1408178
7	49	343	.6457513	.9129312	2	51 84	373 248	.4852814	.1601676
8	64	512	.8284271	2.0000000	3	53 29	389 017	.5440037	.1793392
9	81	729	3.0000000	.0800837	4	54 76	405 224	.6023253	.1983364
10	1 00	1 000	3.1622777	2.1544347	75	56 25	421 875	8.6602540	4.2171633
11	1 21	1 331	.3166248	.2239801	6	57 76	438 976	.7177979	.2358236
12	1 44	1 728	.4641016	.2894286	7	59 29	456 533	.7749644	.2543210
13	1 69	2 197	.6055513	.3513347	8	60 84	474 552	.8317609	.2726586
14	1 96	2 744	.7416574	.4101422	9	62 41	493 039	.8881944	2908404
15	2 25	3 375	3.8729833	2.4662121	80	64 00	512 000	8.9442719	4.3088695
16	2 56	4 096	4.0000000	.5198421	1	65 61	531 441	9.0000000	.3267487
17	2 89	4 913	.1231056	.5712816	2	67 24	551 368	.0553851	.3444815
18	3 24	5 832	.2426407	.6207414	3	68 89	571 787	.1104336	.3620707
19	3 61	6 859	.3588989	.6684016	4	70 56	592 704	.1651514	.3795191
20	4 00	8 000	4.4721360	2.7144177	85	72 25	614 125	9.2195445	4.3968296
1	4 41	9 261	.5825757	2.7589243	6	73 96	636 056	.2736185	.4140049
2	4 84	10 648	.6904158	.8020393	7	75 69	658 503	.3273791	.4310476
3	5 29	12 167	7958315	.8438670	8	77 44	681 472	.3808315	.4479602
4	5 76	13 824	.8989795	.8844991	9	79 21	704 969	.4339811	.4647451
25	6 25	15 625	5.0000000	2.9240177	90	81 00	729 000	9.4868330	4.4814047
6	6 76	17 576	.0990195	.9624960	1	82 81	753 571	.5393920	.4979414
7	7 29	19 683	.1961524	3.0000000	2	84 64	778 688	.5916630	.5143574
8	7 84	21 952	.2915026	.0365889	3	86 49	804 357	.6436508	.5306549
9	8 41	24 389	.3851648	.0723168	4	88 36	830 584	.6953597	.5468359
30	9 00	27 000	5.4772256	3.1072325	95	90 25	857 375	9.7467943	4.5629026
1	9 61	29 791	.5677644	.1413806	6	92 16	884 736	.7979590	.5788570
2	10 24	32 768	.6568542	.1748021	7	94 09	912 673	.8488578	.5947009
3	10 89	35 937	.7445626	.2075343	8	96 04	941 192	.8994949	.6104363
4	11 56	39 304	.8309519	.2396118	9	98 01	970 299	.9498744	.6260650
35	12 25	42 875	5.9160798	3.2710663	100	1 00 00	1 000 000	10.0000000	4.6415888
6	12 96	46 656	6.0000000	.3019272	1	1 02 01	1 030 301	.0498756	.6570095
7	13 69	50 653	.0827625	.3322218	2	1 04 04	1 061 208	.0995049	.6723287
8	14 44	54 872	.1644140	.3619754	3	1 06 09	1 092 927	.1488916	.6875482
9	15 21	59 319	2449980	.3912114	4	1 08 16	1 124 864	.1980390	.7026694
40	16 00	64 000	6.3245553	3.4199519	105	1 10 25	1 157 625	10.2469508	4.7176940
1	16 81	68 921	.4031242	.4482172	6	1 12 36	1 191 016	.2956301	.7326235
2	17 64	74 088	.4807407	.4760266	7	1 14 49	1 225 043	.3440804	.7474594
3	18 49	79 507	.5574385	.5033981	8	1 16 64	1 259 712	.3923048	.7622032
4	19 36	85 184	.6332496	.5303483	9	1 18 81	1 295 029	.4403065	.7768562
45	20 25	91 125	6.7082039	3.5568933	110	1 21 00	1 331 000	10,4880885	4.7914199
6	21 16	97 336	.7823300	.5830479	11	1 23 21	1 367 631	.5356538	.8058955
7	22 09	103 823	.8556546	.6088261	12	1 25 44	1 404 928	.5830052	.8202845
8	23 04	110 592	.9282032	.6342411	13	1 27 69	1 442 897	.6301458	.8345881
9	24 01	117 649	7.0000000	.6593057	14	1 29 96	1 481 544	.6770783	.8488076
50	25 00	125 000	7.0710678	3.6840314	115	1 32 25	1 520 875	10.7238053	4.8629442
1	26 01	132 651	.1414284	.7084298	16	1 34 56	1 560 896	7703296	.8769990
2	27 04	140 608	.2111026	.7325111	17	1 36 89	1 601 613	8166538	.8909732
3	28 09	148 877	.2801099	.7562858	18	1 39 24	1 643 032	.8627805	.9048681
4	29 16	157 464	.3484692	.7797631	19	1 41 61	1 685 159	.9087121	.9186847
55	30 25	166 375	7.4161985	3.8029525	120	1 44 00	1 728 000	10.9544512	4.9324242
6	31 36	175 616	.4833148	.8258624	1	1 46 41	1 771 561	11.0000000	9460874
7	32 49	185 193	.5498344	.8485011	2	1 48 84	1 815 848	.0453610	.9596757
8	33 64	195 112	.6157731	.8708766	3	1 51 29	1 860 867	.0905365	.9731898
9	34 81	205 379	.6811457	.8929965	4	1 53 76	1 906 624	.1355287	.9866310
60	36 00	216 000	7.7459667	3.9148676	125	1 56 25	1 953 125	11.1803399	5.0000000
1	37 21	226 981	.8102497	.9364972	6	1 58 76	2 000 376	.2249722	.0132979
2	38 44	238 328	.8740079	.9578915	7	1 61 29	2 048 383	.2694277	.0265257
3	39 69	250 047	9372539	.9790571	8	1 63 84	2 097 152	3137085	.0396842
4	40 96	262 144	8.0000000	4.0000000	9	1 66 41	2 146 689	.3578167	.0527743
65	42 25	274 625	8.0622577	4.0207256	130	1 69 00	2 197 000	11.4017543	5.0657970

TABLE III

SQUARES, CUBES, SQUARE ROOTS, CUBE ROOTS, OF NUMBERS—*Continued*

No.	Square	Cube.	Sq. Rt.	Cu. Rt.	No.	Square	Cube.	Sq. Rt.	Cu. Rt.
130	1 69 00	2 197 000	11.4017543	5.0657970	195	3 80 25	7 414 875	13.9642400	5.7988900
1	1 71 61	2 248 091	.4455231	.0787531	6	3 84 16	7 529 536	14.0000000	.8087857
2	1 74 24	2 299 968	.4891253	.0916434	7	3 88 09	7 645 373	.0356688	.8186479
3	1 76 89	2 352 637	.5325626	.1044687	8	3 92 04	7 762 392	.0712473	.8284767
4	1 79 56	2 406 104	.5758369	.1172299	9	3 96 01	7 880 599	.1067360	.8382725
135	1 82 25	2 460 375	11.6189500	5.1299278	200	4 00 00	8 000 000	14.1421356	5.8480355
6	1 84 96	2 515 456	.6619038	.1425632	1	4 04 01	8 120 601	.1774469	.8577660
7	1 87 69	2 571 353	.7046999	.1551367	2	4 08 04	8 242 408	.2126704	.8674643
8	1 90 44	2 628 072	.7473401	.1676493	3	4 12 09	8 365 427	.2478068	.8771307
9	1 93 21	2 685 619	.7898261	.1801015	4	4 16 16	8 489 664	.2828569	.8867653
140	1 96 00	2 744 000	11.8321596	5.1924941	205	4 20 25	8 615 125	14.3178211	5.8963685
1	1 98 81	2 803 221	.8743422	.2048279	6	4 24 36	8 741 816	.3527001	.9059406
2	2 01 64	2 863 288	.9163753	.2171034	7	4 28 49	8 869 743	.3874946	.9154817
3	2 04 49	2 924 207	.9582607	.2293215	8	4 32 64	8 998 912	.4222051	.9249921
4	2 07 36	2 985 984	12.0000000	.2414828	9	4 36 81	9 129 329	.4568323	.9344721
145	2 10 25	3 048 625	12.0415946	5.2535879	210	4 41 00	9 261 000	14.4913767	5.9439220
6	2 13 16	3 112 136	.0830460	.2656374	11	4 45 21	9 393 931	.5258390	.9533418
7	2 16 09	3 176 523	.1243557	.2776321	12	4 49 44	9 528 128	.5602198	.9627320
8	2 19 04	3 241 792	.1655251	.2895725	13	4 53 69	9 663 597	.5945195	.9720926
9	2 22 01	3 307 949	.2065556	.3014592	14	4 57 96	9 800 344	.6287388	.9814240
150	2 25 00	3 375 000	12.2474487	5.3132928	215	4 62 25	9 938 375	14.6628783	5.9907264
1	2 28 01	3 442 951	.2882057	.3250740	16	4 66 56	10 077 696	.6969385	6.0000000
2	2 31 04	3 511 808	.3288280	.3368033	17	4 70 89	10 218 313	.7309199	.0092450
3	2 34 09	3 581 577	.3693169	.3484812	18	4 75 24	10 360 232	.7648231	.0184617
4	2 37 16	3 652 264	.4096736	.3601084	19	4 79 61	10 503 459	.7986486	.0276502
155	2 40 25	3 723 875	12.4498996	5.3716854	220	4 84 00	10 648 000	14.8323970	6.0368107
6	2 43 36	3 796 416	.4899960	.3832126	1	4 88 41	10 793 861	.8660687	.0459435
7	2 46 49	3 869 893	.5299641	.3946907	2	4 92 84	10 941 048	.8996644	.0550489
8	2 49 64	3 944 312	.5698051	.4061202	3	4 97 29	11 089 567	.9331845	.0641270
9	2 52 81	4 019 679	.6095202	.4175015	4	5 01 76	11 239 424	.9666295	.0731779
160	2 56 00	4 096 000	12.6491106	5.4288352	225	5 06 25	11 390 625	15.0000000	6.0822020
1	2 59 21	4 173 281	.6885775	.4401218	6	5 10 76	11 543 176	.0332964	.0911994
2	2 62 44	4 251 528	.7279221	.4513618	7	5 15 29	11 697 083	.0665192	.1001702
3	2 65 69	4 330 747	.7671453	.4625556	8	5 19 84	11 852 352	.0996689	.1091147
4	2 68 96	4 410 944	.8062485	.4737037	9	5 24 41	12 008 989	.1327460	.1180332
165	2 72 25	4 492 125	12.8452326	5.4848066	230	5 29 00	12 167 000	15.1657509	6.1269257
6	2 75 56	4 574 296	.8840987	.4958647	1	5 33 61	12 326 391	.1986842	.1357924
7	2 78 89	4 657 463	.9228480	.5068784	2	5 38 24	12 487 168	.2315462	.1446337
8	2 82 24	4 741 632	.9614814	.5178484	3	5 42 89	12 649 337	.2643375	.1534495
9	2 85 61	4 826 809	13.0000000	.5287748	4	5 47 56	12 812 904	.2970585	.1622401
170	2 89 00	4 913 000	13.0384048	5.5396583	235	5 52 25	12 977 875	15.3297097	6.1710058
1	2 92 41	5 000 211	.0766968	.5504991	6	5 56 96	13 144 256	.3622915	.1797466
2	2 95 84	5 088 448	.1148770	.5612978	7	5 61 69	13 312 053	.3948043	.1884628
3	2 99 29	5 177 717	.1529464	.5720546	8	5 66 44	13 481 272	.4272486	.1971544
4	3 02 76	5 268 024	.1909060	.5827702	9	5 71 21	13 651 919	.4596248	.2058218
175	3 06 25	5 359 375	13.2287566	5.5934447	240	5 76 00	13 824 000	15.4919334	6.2144650
6	3 09 76	5 451 776	.2664992	.6040787	1	5 80 81	13 997 521	.5241747	.2230843
7	3 13 29	5 545 233	.3041347	.6146724	2	5 85 64	14 172 488	.5563492	.2316797
8	3 16 84	5 639 752	.3416641	.6252263	3	5 90 49	14 348 907	.5884573	.2402515
9	3 20 41	5 735 339	.3790882	.6357408	4	5 95 36	14 526 784	.6204994	.2487998
180	3 24 00	5 832 000	13.4164079	5.6462162	245	6 00 25	14 706 125	15.6524758	6.2573248
1	3 27 61	5 929 741	.4536240	.6566528	6	6 05 16	14 886 936	.6843871	.2658266
2	3 31 24	6 028 568	.4907376	.6670511	7	6 10 09	15 069 223	.7162336	.2743054
3	3 34 89	6 128 487	.5277493	.6774114	8	6 15 04	15 252 992	.7480157	.2827613
4	3 38 56	6 229 504	.5646600	.6877340	9	6 20 01	15 438 249	.7797338	.2911946
185	3 42 25	6 331 625	13.6014705	5.6980192	250	6 25 00	15 625 000	15.8113883	6.2996053
6	3 45 96	6 434 856	.6381817	.7082675	1	6 30 01	15 813 251	.8429795	.3079935
7	3 49 69	6 539 203	.6747943	.7184791	2	6 35 04	16 003 008	.8745079	.3163596
8	3 53 44	6 644 672	.7113092	.7286543	3	6 40 09	16 194 277	.9059737	.3247035
9	3 57 21	6 751 269	.7477271	.7387936	4	6 45 16	16 387 064	.9373775	.3330256
190	3 61 00	6 859 000	13.7840488	5.7488971	255	6 50 25	16 581 375	15.9687194	6.3413257
1	3 64 81	6 967 871	.8202750	.7589652	6	6 55 36	16 777 216	16.0000000	.3496042
2	3 68 64	7 077 888	.8564065	.7689982	7	6 60 49	16 974 593	.0312195	.3578611
3	3 72 49	7 189 057	.8924440	.7789966	8	6 65 64	17 173 512	.0623784	.3660968
4	3 76 36	7 301 384	.9283883	.7889604	9	6 70 81	17 373 979	.0934769	.3743111
195	3 80 25	7 414 875	13.9642400	5.7988900	260	6 76 00	17 576 000	16.1245155	6.3825043

TABLE III

Squares, Cubes, Square Roots, Cube Roots, of Numbers—*Continued*

No.	Square	Cube.	Sq. Rt.	Cu. Rt.	No.	Square	Cube.	Sq. Rt.	Cu. Rt.
260	6 76 00	17 576 000	16.1245155	6.3825043	325	10 56 25	34 328 125	18.0277564	6.8753443
1	6 81 21	17 779 581	.1554944	.3906765	6	10 62 76	34 645 976	.0554701	.8823888
2	6 86 44	17 984 728	.1864141	.3988279	7	10 69 29	34 965 783	.0831413	.8894188
3	6 91 69	18 191 447	.2172747	.4069585	8	10 75 84	35 287 552	.1107703	.8964345
4	6 96 96	18 399 744	.2480768	.4150687	9	10 82 41	35 611 289	.1383571	.9034359
265	7 02 25	18 609 625	16.2788206	6.4231583	330	10 89 00	35 937 000	18.1659021	6.9104232
6	7 07 56	18 821 096	.3095064	.4312276	1	10 95 61	36 264 691	.1934054	.9173964
7	7 12 89	19 034 163	.3401346	.4392767	2	11 02 24	36 594 368	.2208672	.9243556
8	7 18 24	19 248 832	.3707055	.4473057	3	11 08 89	36 926 037	.2482876	.9313008
9	7 23 61	19 465 109	.4012195	.4553148	4	11 15 56	37 259 704	.2756669	.9382321
270	7 29 00	19 683 000	16.4316767	6.4633041	335	11 22 25	37 595 375	18.3030052	6.9451496
1	7 34 41	19 902 511	.4620776	.4712736	6	11 28 96	37 933 056	.3303028	.9520533
2	7 39 84	20 123 648	.4924225	.4792236	7	11 35 69	38 272 753	.3575598	.9589434
3	7 45 29	20 346 417	.5227116	.4871541	8	11 42 44	38 614 472	.3847763	.9658198
4	7 50 76	20 570 824	.5529454	.4950653	9	11 49 21	38 958 219	.4119526	.9726826
275	7 56 25	20 796 875	16.5831240	6.5029572	340	11 56 00	39 304 000	18.4390889	6.9795321
6	7 61 76	21 024 576	.6132477	.5108300	1	11 62 81	39 651 821	.4661853	.9863681
7	7 67 29	21 253 933	.6433170	.5186839	2	11 69 64	40 001 688	.4932420	.9931906
8	7 72 84	21 484 952	.6733320	.5265189	3	11 76 49	40 353 607	.5202592	7.0000000
9	7 78 41	21 717 639	.7032931	.5343351	4	11 83 36	40 707 584	.5472370	.0067962
280	7 84 00	21 952 000	16.7332005	6.5421326	345	11 90 25	41 063 625	18.5741756	7.0135791
1	7 89 61	22 188 041	.7630546	.5499116	6	11 97 16	41 421 736	.6010752	.0203490
2	7 95 24	22 425 768	.7928556	.5576722	7	12 04 09	41 781 923	.6279360	.0271058
3	8 00 89	22 665 187	.8226038	.5654144	8	12 11 04	42 144 192	.6547581	.0338497
4	8 06 56	22 906 304	.8522995	.5731385	9	12 18 01	42 508 549	.6815417	.0405806
285	8 12 25	23 149 125	16.8819430	6.5808443	350	12 25 00	42 875 000	18.7082869	7.0472987
6	8 17 96	23 393 656	.9115345	.5885323	1	12 32 01	43 243 551	.7349940	.0540041
7	8 23 69	23 639 903	.9410743	.5962023	2	12 39 04	43 614 208	.7616630	.0606967
8	8 29 44	23 887 872	.9705627	.6038545	3	12 46 09	43 986 977	.7882942	.0673767
9	8 35 21	24 137 569	17.0000000	.6114890	4	12 53 16	44 361 864	.8148877	.0740440
290	8 41 00	24 389 000	17.0293864	6.6191060	355	12 60 25	44 738 875	18.8414437	7.0806988
1	8 46 81	24 642 171	.0587221	.6267054	6	12 67 36	45 118 016	.8679623	.0873411
2	8 52 64	24 897 088	.0880075	.6342874	7	12 74 49	45 499 293	.8944436	.0939709
3	8 58 49	25 153 757	.1172428	.6418522	8	12 81 64	45 882 712	.9208879	.1005885
4	8 64 36	25 412 184	.1464282	.6493998	9	12 88 81	46 268 279	.9472953	.1071937
295	8 70 25	25 672 375	17.1755640	6.6569302	360	12 96 00	46 656 000	18.9736660	7.1137866
6	8 76 16	25 934 336	2046505	.6644437	1	13 03 21	47 045 881	19,0000000	.1203674
7	8 82 09	26 198 073	.2336879	.6719403	2	13 10 44	47 437 928	.0262976	.1269300
8	8 88 04	26 463 592	.2626765	.6794200	3	13 17 69	47 832 147	.0525589	.1334925
9	8 94 01	26 730 899	.2916165	.6868831	4	13 24 96	48 228 544	.0787840	.1400370
300	9 00 00	27 000 000	17.3205081	6.6943295	365	13 32 25	48 627 125	19.1049732	7.0465695
1	9 06 01	27 270 901	.3493516	.7017593	6	13 39 56	49 027 896	.1311265	.1530901
2	9 12 04	27 543 608	.3781472	.7091729	7	13 46 89	49 430 863	.1572441	.1595988
3	9 18 09	27 818 127	.4068952	.7165700	8	13 54 24	49 836 032	.1833261	.1660957
4	9 24 16	28 094 464	.4355958	.7239508	9	13 61 61	50 243 409	.2093727	.1725809
305	9 30 25	28 372 625	17.4642492	6.7313155	370	13 69 00	50 653 000	19.2353841	7.1790544
6	9 36 36	28 652 616	.4928557	.7386641	1	13 76 41	51 064 811	.2613603	.1855162
7	9 42 49	28 934 443	.5214155	.7459967	2	13 83 84	51 478 848	.2873015	.1919663
8	9 48 64	29 218 112	.5499288	.7533134	3	13 91 29	51 895 117	.3132079	.1984050
9	9 54 81	29 503 629	.5783958	.7606143	4	13 98 76	52 313 624	.3390796	.2048322
310	9 61 00	29 791 000	17.6068169	6.7678995	375	14 06 25	52 734 375	19.3649167	7.2112479
11	9 67 21	30 080 231	.6351921	.7751690	6	14 13 76	53 157 376	.3907194	.2176522
12	9 73 44	30 371 328	.6635217	.7824229	7	14 21 29	53 582 633	.4164878	.2240450
13	9 79 69	30 664 297	.6918060	.7896613	8	14 28 84	54 010 152	.4422221	.2304268
14	9 85 96	30 959 144	.7200451	.7968844	9	14 36 41	54 439 939	.4679223	.2367972
315	9 92 25	31 255 875	17.7482393	6.8040921	380	14 44 00	54 872 000	19.4935887	7.2431565
16	9 98 56	31 554 496	.7763888	.8112847	1	14 51 61	55 306 341	.5192213	.2495045
17	10 04 89	31 855 013	.8044938	.8184620	2	14 59 24	55 742 968	.5448203	.2558415
18	10 11 24	32 157 432	.8325545	.8256242	3	14 66 89	56 181 887	.5703858	.2621675
19	10 17 61	32 461 759	.8605711	.8327714	4	14 74 56	56 623 104	.5959179	.2684824
320	10 24 00	32 768 000	17.8885438	6.8399037	385	14 82 25	57 066 625	19 6214169	7.2747864
1	10 30 41	33 076 161	.9184729	.8470213	6	14 89 96	57 512 456	.6468827	.2810794
2	10 36 84	33 386 248	.9443584	.8541240	7	14 97 69	57 960 603	.6723156	.2873617
3	10 43 29	33 698 267	.9722008	.8612120	8	15 05 44	58 411 072	.6977156	.2936330
4	10 49 76	34 012 224	18.0000000	.8682855	9	15 13 21	58 863 869	.7230829	.2998936
325	10 56 25	34 328 125	18.0277564	6.8753443	390	15 21 00	59 319 000	19.7484177	7.3061436

TABLE III

Squares, Cubes, Square Roots, Cube Roots, of Numbers—*Continued*

No.	Square	Cube.	Sq. Rt.	Cu. Rt.	No.	Square	Cube.	Sq. Rt.	Cu. Rt.
390	15 21 00	59 319 000	19.7484177	7.3061436	455	20 70 25	94 196 375	21.3307290	7.6913717
1	15 28 81	59 776 471	.7737199	.3123828	6	20 79 36	94 818 816	.3541565	.6970023
2	15 36 64	60 236 288	.7989899	.3186114	7	20 88 49	95 443 993	.3775583	.7026246
3	15 44 49	60 698 457	.8242276	.3248295	8	20 97 64	96 071 912	.4009346	.7082388
4	15 52 36	61 162 984	.8494332	.3310369	9	21 06 81	96 702 579	.4242853	.7138448
395	15 60 25	61 629 875	19.8746069	7.3372339	460	21 16 00	97 336 000	21.4476106	7.7194426
6	15 68 16	82 099 136	.8997487	.3434205	1	21 25 21	97 972 181	.4709106	.7250325
7	15 76 09	62 570 773	.9248588	.3495966	2	21 34 44	98 611 128	.4941853	.7306141
8	15 84 04	63 044 792	.9499373	.3557624	3	21 43 69	99 252 847	.5174348	.7361877
9	15 92 01	63 521 199	.9749844	.3619178	4	21 52 96	99 897 344	.5406592	.7417532
400	16 00 00	64 000 000	20.0000000	7.3680630	465	21 62 25	100 544 625	21.5638587	7.7473109
1	16 08 01	64 481 201	.0249844	.3741979	6	21 71 56	101 194 696	.5870331	.7528606
2	16 16 04	64 964 808	.0499377	.3803227	7	21 80 89	101 847 563	.6101828	.7584023
3	16 24 09	65 450 827	.0748599	.3864373	8	21 90 24	102 503 232	.6333077	.7639361
4	16 32 16	65 939 264	.0997512	.3925418	9	21 99 61	103 161 709	.6564078	.7694620
405	16 40 25	66 430 125	20.1246118	7.3986363	470	22 09 00	103 823 000	21.6794834	7.7749801
6	16 48 36	66 923 416	.1494417	.4047206	1	22 18 41	104 487 111	.7025344	.7804904
7	16 56 49	67 419 143	.1742410	.4107950	2	22 27 84	105 154 048	.7255610	.7859928
8	16 64 64	67 917 312	.1990099	.4168595	3	22 37 29	105 823 817	.7485632	.7914875
9	16 72 81	68 417 929	.2237484	.4229142	4	22 46 76	106 496 424	.7715411	.7969745
410	16 81 00	68 921 000	20.2484567	7.4289589	475	22 56 25	107 171 875	21.7944947	7.8024538
11	16 89 21	69 426 531	.2731349	.4349938	6	22 65 76	107 850 176	.8174242	.8079254
12	16 97 44	69 934 528	.2977831	.4410189	7	22 75 29	108 531 333	.8403297	.8133892
13	17 05 69	70 444 997	.3224014	.4470342	8	22 84 84	109 215 352	.8632111	.8188456
14	17 13 96	70 957 944	.3469899	.4530399	9	22 94 41	109 902 239	.8860686	.8242942
415	17 22 25	71 473 375	20.3715488	7.4590359	480	23 04 00	110 592 000	21.9089023	7.8297353
16	17 30 56	71 991 296	.3960781	.4650223	1	23 13 61	111 284 641	.9317122	.8351688
17	17 38 89	92 511 713	.4205779	.4709991	2	23 23 24	111 980 168	.9544984	.8405949
18	17 47 24	73 034 632	.4450483	.4769664	3	23 32 89	112 678 587	.9772610	.8460134
19	17 55 61	73 560 059	.4694895	.4829242	4	23 42 56	113 379 904	22.0000000	.8514244
420	17 64 00	74 088 000	20.4939015	7.4888724	485	23 52 25	114 084 125	22.0227155	7.8568281
1	17 72 41	74 618 461	.5182845	.4948113	6	23 61 96	114 791 256	.0454077	.8622242
2	17 80 84	75 151 448	.5426386	.5007406	7	23 71 69	115 501 303	.0680765	.8676130
3	17 89 29	75 686 967	.5669638	.5066607	8	23 81 44	116 214 272	.0907220	.8729944
4	17 97 76	76 225 024	.5912603	.5125715	9	23 91 21	116 930 169	.1133444	.8783684
425	18 06 25	76 765 625	20.6155281	7.5184730	490	24 01 00	117 649 000	22.1359436	7.8837352
6	18 14 76	77 308 776	.6397674	.5243652	1	24 10 81	118 370 771	.1585198	.8890946
7	18 23 29	77 854 483	.6639783	.5302482	2	24 20 64	119 095 488	.1810730	.8944468
8	18 31 84	78 402 752	.6881609	.5361221	3	24 30 49	119 823 157	.2036033	.8997917
9	18 40 41	78 953 589	.7123152	.5419867	4	24 40 36	120 553 784	.2261108	.9051294
430	18 49 00	79 507 000	20.7364414	7.5478423	495	24 50 25	121 287 375	22.2485955	7.9104599
1	18 57 61	80 062 991	.7605395	.5536888	6	24 60 16	122 023 936	.2710575	.9157832
2	18 66 24	80 621 568	.7846097	.5595263	7	24 70 09	122 763 473	.2934968	.9210994
3	18 74 89	81 182 737	.8086520	.5653548	8	24 80 04	123 505 992	.3159136	.9264085
4	18 83 56	81 746 504	.8326667	.5711743	9	24 90 01	124 251 499	.3383079	.9317104
435	18 92 25	82 312 875	20.8566536	7.5769849	500	25 00 00	125 000 000	22.3606798	7.9370053
6	19 00 96	82 881 856	.8806130	.5827865	1	25 10 01	125 751 501	.3830293	.9422931
7	19 09 69	83 453 453	.9045450	.5885793	2	25 20 04	126 506 008	.4053565	.9475739
8	19 18 44	84 027 672	.9284495	.5943633	3	25 30 09	127 263 527	.4276615	.9528477
9	19 27 21	84 604 519	.9523268	.6001385	4	25 40 16	128 024 064	.4499443	.9581144
440	19 36 00	35 184 000	20.9761770	7.6059049	505	25 50 25	128 787 625	22.4722051	7.9633743
1	19 44 81	85 766 121	21.0000000	.6116626	6	25 60 36	129 554 216	.4944438	.9686271
2	19 53 64	86 350 888	.0237960	.6174116	7	25 70 49	130 323 843	.5166605	.9738731
3	19 62 49	86 938 307	.0475652	.6231519	8	25 80 64	131 096 512	.5388553	.9791122
4	19 71 36	87 528 384	.0713075	.6288837	9	25 90 81	131 872 229	.5610283	.9843444
445	19 80 25	88 121 125	21.0950231	7.6346067	510	26 01 00	132 651 000	22.5831796	7.7895697
6	19 89 16	88 716 536	.1187121	.6403213	11	26 11 21	133 432 831	.6053091	.9947883
7	19 98 09	89 314 623	.1423745	.6460272	12	26 21 44	134 217 728	.6274170	8.0000000
8	20 07 04	89 915 392	.1660105	.6517247	13	26 31 69	135 005 697	.6495033	.0052049
9	20 16 01	90 518 849	.1896201	.6574138	14	26 41 96	135 796 744	.6715681	.0104032
450	20 25 00	91 125 000	21.2132034	7.6630943	515	26 52 25	136 590 875	22.6936114	8.0155946
1	20 34 01	91 733 851	.2367606	.6687665	16	26 62 56	137 388 096	.7156334	.0207794
2	20 43 04	92 345 408	.2602916	.6744303	17	26 72 89	138 188 413	.7376340	.0259574
3	20 52 09	92 959 677	.2837967	.6800857	18	26 83 24	138 991 832	.7596134	.0311287
4	20 61 16	93 576 664	.3072758	.6857328	19	26 93 61	139 798 359	.7815715	.0362935
455	20 70 25	94 196 375	21.3307290	.6913717	520	27 04 00	140 608 000	22.8035085	8.0414515

TABLE III

Squares, Cubes, Square Roots, Cube Roots, of Numbers—*Continued*

No.	Square	Cube.	Sq. Rt.	Cu. Rt.	No.	Square	Cube.	Sq. Rt.	Cu. Rt.
520	27 04 00	140 608 000	22.8035085	8.0414515	585	34 22 25	200 201 625	24.1867732	8.3634466
1	27 14 41	141 420 761	.8254244	.0466030	6	34 33 96	201 230 056	.2074369	.3682095
2	27 24 84	142 236 648	.8473193	.0517479	7	34 45 69	202 262 003	2280829	.3729668
3	27 35 29	143 055 667	.8691933	.0568862	8	34 57 44	203 297 472	.2487113	.3777188
4	27 45 76	143 877 824	.8910463	.0620180	9	34 69 21	204 336 469	.2693222	.3824653
525	27 56 25	144 703 125	22.9128785	8.0671432	590	34 81 00	205 379 000	24.2899156	8.3872065
6	27 66 76	145 531 576	.9346899	.0722620	1	34 92 81	206 425 071	.3104916	.3919423
7	27 77 29	146 363 183	.9564806	.0773743	2	35 04 64	207 474 688	.3310501	.3966729
8	27 87 84	147 197 952	.9782506	.0824800	3	35 16 49	208 527 857	.3515913	.4013981
9	27 98 41	148 035 889	23.0000000	.0875794	4	35 28 36	209 584 584	.3721152	.4061180
530	28 09 00	148 877 000	23.0217289	8.0926723	595	35 40 25	210 644 875	24.3926218	8.4108326
1	28 19 61	149 721 291	.0434372	0977589	6	35 52 16	211 708 736	.4131112	.4155419
2	28 30 24	150 568 768	.0651252	.1028390	7	35 64 09	212 776 173	.4335834	.4202460
3	28 40 89	151 419 437	.0867928	.1079128	8	35 76 04	213 847 192	.4540385	.4249448
4	28 51 56	152 273 304	.1084400	.1129803	9	35 88 01	214 921 799	.4744765	.4296383
535	28 62 25	153 130 375	23.1300670	8 1180414	600	36 00 00	216 000 000	24.4948974	8.4343267
6	28 72 96	153 990 656	.1516738	.1230962	1	36 12 01	217 081 801	.5153013	.4390098
7	28 83 69	154 854 153	.1732605	.1281447	2	36 24 04	218 167 208	.5356883	.4436877
8	28 94 44	155 720 872	.1948270	.1331870	3	36 36 09	219 256 227	.5560583	.4483605
9	29 05 21	156 590 819	.2163735	.1382230	4	36 48 16	220 348 864	.5764115	.4530281
540	29 16 00	157 464 000	23.2379001	8.1432529	605	36 60 25	221 445 125	24.5967478	8.4576906
1	29 26 81	158 340 421	.2594067	.1482765	6	36 72 36	222 545 016	.6170673	.4623479
2	29 37 64	159 220 088	.2808935	.1532939	7	36 84 49	223 648 543	.6373700	.4670000
3	29 48 49	160 103 007	.3023604	.1583051	8	36 96 64	224 755 712	.6576560	.4716471
4	29 59 36	160 989 184	.3238076	.1633102	9	37 08 81	225 866 529	.6779254	.4762892
545	29 70 25	161 878 625	23.3452351	8.1683092	610	37 21 00	226 981 000	24.6981781	8.4809261
6	29 81 16	162 771 336	.3666429	.1733020	11	37 33 21	228 099 131	.7184142	.4855579
7	29 92 09	163 667 323	.3880311	.1782888	12	37 45 44	229 220 928	.7386338	.4901848
8	30 03 04	164 566 592	.4093998	.1832695	13	37 57 69	230 346 397	.7588368	.4948065
9	30 14 01	165 469 149	.4307490	.1882441	14	37 69 96	231 475 544	.7790234	.4994233
550	30 25 00	166 375 000	23.4520788	8.1932127	615	37 82 25	232 608 375	24.7991935	8.5040350
1	30 36 01	167 284 151	.4733892	.1981753	16	37 94 56	233 744 896	.8193473	.5086417
2	30 47 04	168 196 608	.4946802	.2031319	17	38 06 89	234 885 113	.8394847	.5132435
3	30 58 09	169 112 377	.5159520	.2080825	18	38 19 24	236 029 032	.8596058	.5178403
4	30 69 16	170 031 464	.5372046	.2130271	19	38 31 61	237 176 659	.8797106	.5224321
555	30 80 25	170 953 875	23.5584380	8.2179657	620	38 44 00	238 328 000	24.8997992	8.5270189
6	30 91 36	171 879 616	.5796522	.2228985	1	38 56 41	239 483 061	.9198716	.5316009
7	31 02 49	172 808 693	.6008474	.2278254	2	38 68 84	240 641 848	.9399278	.5361780
8	31 13 64	173 741 112	.6220236	.2327463	3	38 81 29	241 804 367	.9599679	.5407501
9	31 24 81	174 676 879	.6431808	.2376614	4	38 93 76	242 970 624	.9799920	.5453173
560	31 36 00	175 616 000	23.6643191	8.2425706	625	39 06 25	244 140 625	25.0000000	8.5498797
1	31 47 21	176 558 481	.6854386	.2474740	6	39 18 76	245 314 376	.0199920	.5544372
2	31 58 44	177 504 328	.7065392	.2523715	7	39 31 29	246 491 883	.0399681	.5589899
3	31 69 69	178 453 547	.7276210	.2572633	8	39 43 84	247 673 152	.0599282	.5635377
4	31 80 96	179 406 144	.7486842	.2621492	9	39 56 41	248 858 189	.0798724	.5680807
565	31 92 25	180 362 125	23.7697286	8.2670294	630	39 69 00	250 047 000	25.0998008	8.5726189
6	32 03 56	181 321 496	.7907545	.2719039	1	39 81 61	251 239 591	.1197134	.5771523
7	32 14 89	182 284 263	.8117618	.2767726	2	39 94 24	252 435 968	.1396102	.5816809
8	32 26 24	183 250 432	.8327506	.2816355	3	40 06 89	253 636 137	.1594913	.5862047
9	32 37 61	184 220 009	.8537209	.2864928	4	40 19 56	254 840 104	.1793566	.5907238
570	32 49 00	185 193 000	23.8746728	8.2913444	635	40 32 25	256 047 875	25.1992063	8.5952380
1	32 60 41	186 169 411	.8956063	.2961903	6	40 44 96	257 259 456	.2190404	.5997476
2	32 71 84	187 149 248	.9165215	.3010304	7	40 57 69	258 474 853	.2388589	.6042525
3	32 83 29	188 132 517	.9374184	.3058651	8	40 70 44	259 694 072	.2586619	.6087526
4	32 94 76	189 119 224	.9582971	.3106941	9	40 83 21	260 917 119	.2784493	.6132480
575	33 06 25	190 109 375	23.9791576	8.3155175	640	40 96 00	262 144 000	25.2982213	8.6177388
6	33 17 76	191 102 976	24.0000000	.3203353	1	41 08 81	263 374 721	.3179778	.6222248
7	33 29 29	192 100 033	.0208243	.3251475	2	41 21 64	264 609 288	.3377189	.6267063
8	33 40 84	193 100 552	.0416306	.3299542	3	41 34 49	265 847 707	.3574447	.6311830
9	33 52 41	194 104 539	.0624188	.3347553	4	41 47 36	267 089 984	.3771551	.6356551
580	33 64 00	195 112 000	24.0831891	8.3395509	645	41 60 25	268 336 125	25.3968502	8.6401226
1	33 75 61	196 122 941	.1039416	.3443410	6	41 73 16	269 586 136	.4165301	.6445855
2	33 87 24	197 137 368	.1246762	.3491256	7	41 86 09	270 840 023	.4361947	.6490437
3	33 98 89	198 155 287	.1453929	.3539047	8	41 99 04	272 097 792	.4558441	.6534974
4	34 10 56	199 176 704	.1660919	.3586784	9	42 12 01	273 359 449	.4754784	.6579465
585	34 22 25	200 201 625	24.1867732	8.3634466	650	42 25 00	274 625 000	25.4950976	8.6623911

TABLE III

SQUARES, CUBES, SQUARE ROOTS, CUBE ROOTS, OF NUMBERS—*Continued*

No.	Square	Cube.	Sq. Rt.	Cu. Rt.	No.	Square	Cube.	Sq. Rt.	Cu. Rt.
650	42 25 00	274 625 000	25.4950976	8.6623911	715	51 12 25	365 525 875	26.7394839	8.9420140
1	42 38 01	275 894 451	.5147016	.6668310	16	51 26 56	367 061 696	.7581763	.9461809
2	42 51 04	277 167 808	.5342907	.6712665	17	51 40 89	368 601 813	.7768557	.9503438
3	42 64 09	278 445 077	.5538647	.6756974	18	51 55 24	370 146 232	.7955220	.9545029
4	42 77 16	279 726 264	.5734237	.6801237	19	51 69 61	371 694 959	.8141754	.9586581
655	42 90 25	281 011 375	25.5929678	8.6845456	720	51 84 00	373 248 000	26.8328157	8.9628095
6	43 03 36	282 300 416	.6124969	.6889630	1	51 98 41	374 805 361	.8514432	.9669570
7	43 16 49	283 593 393	.6320112	.6933759	2	52 12 84	376 367 048	.8700577	.9711007
8	43 29 64	284 890 312	.6515107	.6977843	3	52 27 29	377 933 067	.8886593	.9752406
9	43 42 81	286 191 179	.6709953	.7021882	4	52 41 76	379 503 424	.9072481	.9793766
660	43 56 00	287 496 000	25.6904652	8.7065877	725	52 56 25	381 078 125	26.9258240	8.9835089
1	43 69 21	288 804 781	.7099203	.7109827	6	52 70 76	382 657 176	.9443872	.9876373
2	43 82 44	290 117 528	.7293607	.7153734	7	52 85 29	384 240 583	.9629375	.9917620
3	43 95 69	291 434 247	.7487864	.7197596	8	52 99 84	385 828 352	.9814751	.9958829
4	44 08 96	292 754 944	.7681975	.7241414	9	53 14 41	387 420 489	27.0000000	9.0000000
665	44 22 25	294 079 625	25.7875939	8.7285187	730	53 29 00	389 017 000	27.0185122	9.0041134
6	44 35 56	295 408 296	.8069758	.7328918	1	53 43 61	390 617 891	.0370117	.0082229
7	44 48 89	296 740 963	.8263431	.7372604	2	53 58 24	392 223 168	.0554985	.0123288
8	44 62 24	298 077 632	.8456960	.7416246	3	53 72 89	393 832 837	.0739727	.0164309
9	44 75 61	299 418 309	.8650343	.7459846	4	53 87 56	395 446 904	.0924344	.0205293
670	44 89 00	300 763 000	25.8843582	8.7503401	735	54 02 25	397 065 375	27.1108834	9.0246239
1	45 02 41	302 111 711	.9036677	.7546913	6	54 16 96	398 688 256	.1293199	.0287149
2	45 15 84	303 464 448	.9229628	.7590383	7	54 31 69	400 315 553	.1477439	.0328021
3	45 29 29	304 821 217	.9422435	.7633809	8	54 46 44	401 947 272	.1661554	.0368857
4	45 42 76	306 182 024	.9615100	.7677192	9	54 61 21	403 583 419	.1845544	.0409655
675	45 56 25	307 546 875	25.9807621	8.7720532	740	54 76 00	405 224 000	27.2029410	9.0450417
6	45 69 76	308 915 776	26.0000000	.7763830	1	54 90 81	406 869 021	.2213152	.0491142
7	45 83 29	310 288 733	.0192237	.7807084	2	55 05 64	408 518 488	.2396769	.0531831
8	45 96 84	311 665 752	.0384331	.7850296	3	55 20 49	410 172 407	.2580263	.0572482
9	46 10 41	313 046 839	.0576284	.7893466	4	55 35 36	411 830 784	.2763634	.0613098
680	46 24 00	314 432 000	26.0768096	8.7936593	745	55 50 25	413 493 625	27.2946881	9.0653677
1	46 37 61	315 821 241	.0959767	.7979679	6	55 65 16	415 160 936	.3130006	.0694220
2	46 51 24	317 214 568	.1151297	.8022721	7	55 80 09	416 832 723	.3313007	.0734726
3	46 64 89	318 611 987	.1342687	.8065722	8	55 95 04	418 508 992	.3495887	.0775197
4	46 78 56	320 013 504	.1533937	.8108681	9	56 10 01	420 189 749	.3678644	.0815631
685	46 92 25	321 419 125	26.1725047	8.8151598	750	56 25 00	421 875 000	27.3861279	9.0856030
6	47 05 96	322 828 856	.1916017	.8194474	1	56 40 01	423 564 751	.4043792	.0896392
7	47 19 69	324 242 703	.2106848	.8237307	2	56 55 04	425 259 008	.4226184	.0936719
8	47 33 44	325 660 672	.2297541	.8280099	3	56 70 09	426 957 777	.4408455	.0977010
9	47 47 21	327 082 769	.2488095	.8322850	4	56 85 16	428 661 064	.4590604	.1017265
690	47 61 00	328 509 000	26.2678511	8.8365559	755	57 00 25	430 368 875	27.4772633	9.1057485
1	47 74 81	329 939 371	.2868789	.8408227	6	57 15 36	432 081 216	.4954542	.1097669
2	47 88 64	331 373 888	.3058929	.8450854	7	57 30 49	433 798 093	.5136330	.1137818
3	48 02 49	332 812 557	.3248932	.8493440	8	57 45 64	435 519 512	.5317998	.1177931
4	48 16 36	334 255 384	.3438797	.8535985	9	57 60 81	437 245 479	.5499546	.1218010
695	48 30 25	335 702 375	26.3628527	8.8578489	760	57 76 00	438 976 000	27.5680975	9.1258053
6	48 44 16	337 153 536	.3818119	.8620952	1	57 91 21	440 711 081	.5862284	.1298061
7	48 58 09	338 608 873	.4007576	.8663375	2	58 06 44	442 450 728	.6043475	.1338034
8	48 72 04	340 068 392	.4196896	.8705757	3	58 21 69	444 194 947	.6224546	.1377971
9	48 86 01	341 532 099	.4386081	.8748099	4	58 36 96	445 943 744	.6405499	.1417874
700	49 00 00	343 000 000	26.4575131	8.8790400	765	58 52 25	447 697 125	27.6586334	9.1457742
1	49 14 01	344 472 101	.4764046	.8832661	6	58 67 56	449 455 096	.6767050	.1497576
2	49 28 04	345 948 408	.4952826	.8874882	7	58 82 89	451 217 663	.6947648	.1537375
3	49 42 09	347 428 927	.5141472	.8917063	8	58 98 24	452 984 832	.7128129	.1577139
4	49 56 16	348 913 664	.5329983	.8959204	9	59 13 61	454 756 609	.7308492	.1616869
705	49 70 25	350 402 625	26.5518361	8.9001304	770	59 29 00	456 533 000	27.7488739	9.1656565
6	49 84 36	351 895 816	.5706605	.9043366	1	59 44 41	458 314 011	.7668868	.1696225
7	49 98 49	353 393 243	.5894716	.9085387	2	59 59 84	460 099 648	.7848880	.1735852
8	50 12 64	354 894 912	.6082694	.9127369	3	59 75 29	461 889 917	.8028775	.1775445
9	50 26 81	356 400 829	6270539	.9169311	4	59 90 76	463 684 824	.8208555	.1815003
710	50 41 00	357 911 000	26.6458252	8.9211214	775	60 06 25	465 484 375	27.8388218	9.1854527
11	50 55 21	359 425 431	.6645833	.9253078	6	60 21 76	467 288 576	.8567766	.1894018
12	50 69 44	360 944 128	.6833281	.9294902	7	60 37 29	469 097 433	.8747197	.1933474
13	50 83 69	362 467 097	.7020598	.9336687	8	60 52 84	470 910 952	.8926514	.1972897
14	50 97 96	363 994 344	.7207784	.9378433	9	60 68 41	472 729 139	.9105715	.2012286
715	51 12 25	365 525 875	26.7394839	8.9420140	780	60 84 00	474 552 000	27.9284801	9.2051641

TABLE III

Squares, Cubes, Square Roots, Cube Roots, of Numbers—*Continued*

No.	Square	Cube.	Sq. Rt.	Cu. Rt.	No.	Square	Cube.	Sq. Rt.	Cu. Rt.
780	60 84 00	474 552 000	27.9284801	9.2051641	845	71 40 25	603 351 125	29.0688837	9.4540719
1	60 99 61	476 379 541	.9463772	.2090962	6	71 57 16	605 495 736	.0860791	.4577999
2	61 15 24	478 211 768	.9642629	.2130250	7	71 74 09	607 645 423	.1032644	.4615249
3	61 30 89	480 048 687	.9821372	.2169505	8	71 91 04	609 800 192	.1204396	.4652470
4	61 46 56	481 890 304	28.0000000	.2208726	9	72 08 01	611 960 049	.1376046	.4689661
785	61 62 25	483 736 625	28.0178515	9.2247914	850	72 25 00	614 125 000	29.1547595	9.4726824
6	61 77 96	485 587 656	.0356915	.2287068	1	72 42 01	616 295 051	.1719043	.4763957
7	61 93 69	487 443 403	.0535203	.2326189	2	72 59 04	618 470 208	.1890390	.4801061
8	62 09 44	489 303 872	.0713377	.2365277	3	72 76 09	620 650 477	.2061637	.4838136
9	62 25 21	491 169 069	.0891438	.2404333	4	72 93 16	622 835 864	.2232784	.4875182
790	62 41 00	493 039 000	28.1069386	9.2443355	855	73 10 25	625 026 375	29.2403830	9.4912200
1	62 56 81	494 913 671	.1247222	.2482344	6	73 27 36	627 222 016	.2574777	.4949188
2	62 72 64	496 793 088	.1424946	.2521300	7	73 44 49	629 422 793	.2745623	.4986147
3	62 88 49	498 677 257	.1602557	.2560224	8	73 61 64	631 628 712	.2916370	.5023078
4	63 04 36	500 566 184	.1780056	.2599114	9	73 78 81	633 839 779	.3087018	.5059980
795	63 20 25	502 459 875	28.1957444	9.2637973	860	73 96 00	636 056 000	29.3257566	9.5096854
6	63 36 16	504 358 336	.2134720	.2676798	1	74 13 21	638 277 381	.3428015	.5133699
7	63 52 09	506 261 573	.2311884	.2715592	2	74 30 44	640 503 928	.3598365	.5170515
8	63 68 04	508 169 592	.2488938	.2754352	3	74 47 69	642 735 647	.3768616	.5207303
9	63 84 01	510 082 399	.2665881	.2793081	4	74 64 96	644 972 544	.3938769	.5244063
800	64 00 00	512 000 000	28.2842712	9.2831777	865	74 82 25	647 214 625	29.4108823	9.5280794
1	64 16 01	513 922 401	.3019434	.2870440	6	74 99 56	649 461 896	.4278779	.5317497
2	64 32 04	515 849 608	.3196045	.2909072	7	75 16 89	651 714 363	.4448637	.5354172
3	64 48 09	517 781 627	.3372546	.2947671	8	75 34 24	653 972 032	.4618397	.5390818
4	64 64 16	519 718 464	.3548938	.2986239	9	75 51 61	656 234 909	.4788059	.5427437
805	64 80 25	521 660 125	28.3725219	9.3024775	870	75 69 00	658 503 000	29.4957624	9.5464027
6	64 96 36	523 606 616	.3901391	.3063278	1	75 86 41	660 776 311	.5127091	.5500589
7	65 12 49	525 557 943	.4077454	.3101750	2	76 03 84	663 054 848	.5296461	.5537123
8	65 28 64	527 514 112	.4253408	.3140190	3	76 21 29	665 338 617	.5465734	.5573630
9	65 44 81	529 475 129	.4429253	.3178599	4	76 38 76	667 627 624	.5634910	.5610108
810	65 61 00	531 441 000	28.4604989	9.3216975	875	76 56 25	669 921 875	29.5803989	9.5646559
11	65 77 21	533 411 731	.4780617	.3255320	6	76 73 76	672 221 376	.5972972	.5682982
12	65 93 44	535 387 328	.4956137	.3293634	7	76 91 29	674 526 133	.6141858	.5719377
13	66 09 69	537 367 797	.5131549	.3331916	8	77 08 84	676 836 152	.6310648	.5755745
14	66 25 96	539 353 144	.5306852	.3370167	9	77 26 41	679 151 439	.6479342	.5792085
815	66 42 25	541 343 375	28.5482048	9.3408386	880	77 44 00	681 472 000	29.6647939	9.5828397
16	66 58 56	543 338 496	.5657137	.3446575	1	77 61 61	683 797 841	.6816442	.5864682
17	66 74 89	545 338 513	.5832119	.3484731	2	77 79 24	686 128 968	.6984848	.5900939
18	66 91 24	547 343 432	.6006993	.3522857	3	77 96 89	688 465 387	.7153159	.5937169
19	67 07 61	549 353 259	.6181760	.3560952	4	78 14 56	690 807 104	.7321375	.5973373
820	67 24 00	551 368 000	28.6356421	9.3599016	885	78 32 25	693 154 125	29.7489496	9.6009548
1	67 40 41	553 387 661	.6530976	.3637049	6	78 49 96	695 506 456	.7657521	.6045696
2	67 56 84	555 412 248	.6705424	.3675051	7	78 67 69	697 864 103	.7825452	.6081817
3	67 73 29	557 441 767	.6879766	.3713022	8	78 85 44	700 227 072	.7993289	.6117911
4	67 89 76	559 476 224	.7054002	.3750963	9	79 03 21	702 595 369	.8161030	.6153977
825	68 06 25	561 515 625	28.7228132	9.3788873	890	79 21 00	704 969 000	29.8328678	9.6190017
6	68 22 76	563 559 976	.7402157	.3826752	1	79 38 81	707 347 971	.8496231	.6226030
7	68 39 29	565 609 283	.7576077	.3864600	2	79 56 64	709 732 288	.8663690	.6262016
8	68 55 84	567 663 552	.7749891	.3902419	3	79 74 49	712 121 957	.8831056	.6297975
9	68 72 41	569 722 789	.7923601	.3940206	4	79 92 36	714 516 984	.8998328	.6333907
830	68 89 00	571 787 000	28.8097206	9.3977964	895	80 10 25	716 917 375	29.9165506	9.6369812
1	69 05 61	573 856 191	.8270706	.4015691	6	80 28 16	719 323 136	.9332591	.6405690
2	69 22 24	575 930 368	.8444102	.4053387	7	80 46 09	721 734 273	.9499583	.6441542
3	69 38 89	578 009 537	.8617394	.4091054	8	80 64 04	724 150 792	.9666481	6477367
4	69 55 56	580 093 704	.8790582	.4128690	9	80 82 01	726 572 699	.9833287	.6513166
835	69 72 25	582 182 875	28.8963666	9 4166297	900	81 00 00	729 000 000	30.0000000	9.6548938
6	69 88 96	584 277 056	.9136646	.4203873	1	81 18 01	731 432 701	.0166620	.6584684
7	70 05 69	586 376 253	.9309523	.4241420	2	81 36 04	733 870 808	.0333148	.6620403
8	70 22 44	588 480 472	.9482297	.4278936	3	81 54 09	736 314 327	.0499584	.6656096
9	70 39 21	590 589 719	.9654967	.4316423	4	81 72 16	738 763 264	.0665928	.6691762
840	70 56 00	592 704 000	28.9827535	9.4353880	905	81 90 25	741 217 625	30.0832179	9.6727403
1	70 72 81	594 823 321	29.0000000	.4391307	6	82 08 36	743 677 416	.0998339	.6763017
2	70 89 64	596 947 688	.0172363	.4428704	7	82 26 49	746 142 643	.1164407	.6798604
3	71 06 49	599 077 107	.0344623	.4466072	8	82 44 64	748 613 312	.1330383	.6834166
4	71 23 36	601 211 584	.0516781	.4503410	9	82 62 81	751 089 429	.1496269	.6869701
845	71 40 25	603 351 125	29.0688837	9.4540719	910	82 81 00	753 571 000	30.1662063	9.6905211

TABLE III

Squares, Cubes, Square Roots, Cube Roots, of Numbers—*Continued*

No.	Square	Cube.	Sq. Rt.	Cu. Rt.	No.	Square.	Cube.	Sq. Rt.	Cu. Rt.
910	82 81 00	753 571 000	30.1662063	9.6905211	975	95 06 25	926 859 375	31.2249900	9.9159624
11	82 99 21	756 058 031	.1827765	.6940694	6	95 25 76	929 714 176	.2409987	.9193513
12	83 17 44	758 550 528	.1993377	.6976151	7	95 45 29	932 574 833	.2569992	.9227379
13	83 35 69	761 048 497	.2158899	.7011583	8	95 64 84	935 441 352	.2729915	.9261222
14	83 53 96	763 551 944	.2324329	.7046989	9	95 84 41	938 313 739	.2889757	.9295042
915	83 72 25	766 060 875	30.2489669	9.7082369	980	96 04 00	941 192 000	01.3049517	9.9328839
16	83 90 56	768 575 296	.2654919	.7117723	1	96 23 61	944 076 141	.3209195	.9362613
17	84 08 89	771 095 213	.2820079	.7153051	2	96 43 24	946 966 168	.3368792	.9396363
18	84 27 24	773 620 632	.2985148	.7188354	3	96 62 89	949 862 087	.3528308	.9430092
19	84 45 61	776 151 559	.3150128	.7223631	4	96 82 56	952 763 904	.3687743	.9463797
920	84 64 00	778 688 000	30.3315018	9.7258883	985	97 02 25	955 671 625	31.3847097	9.9497479
1	84 82 41	781 229 961	.3479818	.7294109	6	97 21 96	958 585 256	.4006369	.9531138
2	85 00 84	783 777 448	.3644529	.7329309	7	97 41 69	961 504 803	.4165561	.9564775
3	85 19 29	786 330 467	.3809151	.7364484	8	97 61 44	964 430 272	.4324673	.9598389
4	85 37 76	788 889 024	.3973683	.7399634	9	97 81 21	967 361 669	.4483704	.9631981
925	85 56 25	791 453 125	30.4138127	9.7434758	990	98 01 00	970 299 000	31.4642654	9.9665549
6	85 74 76	794 022 776	.4302481	.7469857	1	98 20 81	973 242 271	.4801525	.9699095
7	85 93 29	796 597 983	.4466747	.7504930	2	98 40 64	976 191 488	.4960315	.9732619
8	86 11 84	799 178 752	.4630924	.7539979	3	98 60 49	979 146 657	.5119025	.9766120
9	86 30 41	801 765 089	.4795013	.7575002	4	98 80 36	982 107 784	.5277655	.9799599
930	86 49 00	804 357 000	30.4959014	9.7610001	995	99 00 25	985 074 875	31.5436206	9.9833055
1	86 67 61	806 954 491	.5122926	.7644974	6	99 20 16	988 047 936	.5594677	.9866488
2	86 86 24	809 557 568	.5286750	.7679922	7	99 40 09	991 026 973	.5753068	.9899900
3	87 04 89	812 166 237	.5450487	.7714845	8	99 60 04	994 011 992	.5911380	.9933289
4	87 23 56	814 780 504	.5614136	.7749743	9	99 80 01	997 002 999	.6069613	.9966656
935	87 42 25	817 400 375	30.5777697	9.7784616	1000	1 00 00 00	1 000 000 000	31.6227766	10.0000000
6	87 60 96	820 025 856	.5941171	.7819466	1	1 00 20 01	1 003 003 001	.6385840	.0033322
7	87 79 69	822 656 953	.6104557	.7854288	2	1 00 40 04	1 006 012 008	.6543836	.0066622
8	87 98 44	825 293 672	.6267857	.7889087	3	1 00 60 09	1 009 027 027	.6701752	.0099899
9	88 17 21	827 936 019	.6431069	.7923861	4	1 00 80 16	1 012 048 064	.6859590	.0133155
940	88 36 00	830 584 000	30.6594194	9.7958611	1005	1 01 00 25	1 015 075 125	31.7017349	10.0166389
1	88 54 81	833 237 621	.6757233	.7993336	6	1 01 20 36	1 018 108 216	.7175030	.0199601
2	88 73 64	835 896 888	.6920185	.8028036	7	1 01 40 49	1 021 147 343	.7332633	.0232791
3	88 92 49	838 561 807	.7083051	.8062711	8	1 01 60 64	1 024 192 512	.7490157	.0265958
4	89 11 36	841 232 384	.7245830	.8097362	9	1 01 80 81	1 027 243 729	.7647603	.0299104
945	89 30 25	843 908 625	30.7408523	9.8131989	1010	1 02 01 00	1 030 301 000	31.7804972	10.0332228
6	89 49 16	846 590 536	.7571130	.8166591	11	1 02 21 21	1 033 364 331	.7962262	.0365330
7	89 68 09	849 278 123	.7733651	.8201169	12	1 02 41 44	1 036 433 728	.8119474	.0398410
8	89 87 04	851 971 392	.7896086	.8235723	13	1 02 61 69	1 039 509 197	.8276609	.0431469
9	90 06 01	854 670 349	.8058436	.8270252	14	1 02 81 96	1 042 590 744	.8433666	.0464506
950	90 25 00	857 375 000	30.8220700	9.8304757	1015	1 03 02 25	1 045 678 375	31.8590646	10.0497521
1	90 44 01	860 085 351	.8382879	.8339238	16	1 03 22 56	1 048 772 096	.8747549	.0530514
2	90 63 04	862 801 408	.8544972	.8373695	17	1 03 42 89	1 051 871 913	.8904374	.0563485
3	90 82 09	865 523 177	.8706981	.8408127	18	1 03 63 24	1 054 977 832	.9061123	.0596435
4	91 01 16	868 250 664	.8868904	.8442536	19	1 03 83 61	1 058 089 859	.9217794	.0629364
955	91 20 25	870 983 875	30.9030743	9.8476920	1020	1 04 04 00	1 061 208 000	31.9374388	10.0662271
6	91 39 36	873 722 816	.9192497	.8511280	21	1 04 24 41	1 064 332 261	.9530906	.0695156
7	91 58 49	876 467 493	.9354166	.8545617	22	1 04 44 84	1 067 462 648	.9687347	.0728020
8	91 77 64	879 217 912	.9515751	.8579929	23	1 04 65 29	1 070 599 167	.9843712	.0760863
9	91 96 81	881 974 079	.9677251	.8614218	24	1 04 85 76	1 073 741 824	32.0000000	.0793684
960	92 16 00	884 736 000	30.9838668	9.8648483	1025	1 05 06 25	1 076 890 625	32.0156212	10.0826484
1	92 35 21	887 503 681	31.0000000	.8682724	26	1 05 26 76	1 080 045 576	.0312348	.0859262
2	92 54 44	890 277 128	.0161248	.8716941	27	1 05 47 29	1 083 206 683	.0468407	.0892019
3	92 73 69	893 056 347	.0322413	.8751135	28	1 05 67 84	1 086 373 952	.0624391	.0924755
4	92 92 96	895 841 344	.0483494	.8785305	29	1 05 88 41	1 089 547 389	.0780298	.0957469
965	93 12 25	898 632 125	31.0644491	9.8819451	1030	1 06 09 00	1 092 727 000	32.0936131	10.0990163
6	93 31 56	901 428 696	.0805405	.8853574	31	1 06 29 61	1 095 912 791	.1091887	.1022835
7	93 50 89	904 231 063	.0966236	.8887673	32	1 06 50 24	1 099 104 768	.1247568	.1055487
8	93 70 24	907 039 232	.1126984	.8921749	33	1 06 70 89	1 102 302 937	.1403173	.1088117
9	93 89 61	909 853 209	.1287648	.8955801	34	1 06 91 56	1 105 507 304	.1558704	.1120726
970	94 09 00	912 673 000	31.1448230	9.8989830	1035	1 07 12 25	1 108 717 875	32.1714159	10.1153314
1	94 28 41	915 498 611	.1608729	.9023835	36	1 07 32 96	1 111 934 656	.1869539	.1185882
2	94 47 84	918 330 048	.1769145	.9057817	37	1 07 53 69	1 115 157 653	.2024844	.1218428
3	94 67 29	921 167 317	.1929479	.9091776	38	1 07 74 44	1 118 386 872	.2180074	.1250953
4	94 86 76	924 010 424	.2089731	.9125712	39	1 07 95 21	1 121 622 319	.2335229	.1283457
975	95 06 25	926 859 375	31.2249900	9.9159624	1040	1 08 16 00	1 124 864 000	32.2490310	10.1315941

TABLE III

Squares, Cubes, Square Roots, Cube Roots, of Numbers—*Continued*

No.	Square	Cube.	Sq. Rt.	Cu. Rt.	No.	Square	Cube.	Sq. Rt.	Cu. Rt.
1040	1 08 16 00	1 124 864 000	32.2490310	10.1315941	1105	1 22 10 25	1 349 232 625	33.2415403	10.3384181
41	1 08 36 81	1 128 111 921	.2645316	.1348403	6	1 22 32 36	1 352 899 016	.2565783	.3415358
42	1 08 57 64	1 131 366 088	.2800248	.1380845	7	1 22 54 49	1 356 572 043	.2716095	.3446517
·43	1 08 78 49	1 134 626 507	.2955105	.1413266	8	1 22 76 64	1 360 251 712	.2866339	.3477657
44	1 08 99 36	1 137 893 184	.3109888	.1445667	9	1 22 98 81	1 363 938 029	.3016516	.3508778
1045	1 09 20 25	1 141 166 125	32.3264598	10.1478047	1110	1 23 21 00	1 367 631 000	33.3166625	10.3539880
46	1 09 41 16	1 144 445 336	.3419233	.1510406	11	1 23 43 21	1 371 330 631	.3316666	.3570964
47	1 09 62 09	1 147 730 823	.3573794	.1542744	12	1 23 65 44	1 375 036 928	.3466640	.3602029
48	1 10 83 04	1 151 022 592	.3728281	.1575062	13	1 23 87 69	1 378 749 897	.3616546	.3633076
49	1 10 04 01	1 154 320 649	.3882695	.1607359	14	1 24 09 96	1 382 469 544	.3766385	.3664103
1050	1 10 25 00	1 157 625 000	32.4037035	10.1639636	1115	1 24 32 25	1 386 195 875	33.3916157	10.3695113
51	1 10 46 01	1 160 935 651	.4191301	.1671893	16	1 24 54 56	1 389 928 896	.4065862	.3726103
52	1 10 67 04	1 164 252 608	.4345495	.1704129	17	1 24 76 89	1 393 668 613	.4215499	.3757076
53	1 10 88 09	1 167 575 877	.4499615	.1736344	18	1 24 99 24	1 397 415 032	.4365070	.3788030
54	1 11 09 16	1 170 905 464	.4653662	.1768539	19	1 25 21 61	1 401 168 159	.4514573	.3818965
1055	1 11 30 25	1 174 241 375	32.4807635	10.1800714	1120	1 25 44 00	1 404 928 000	33.4664011	10.3849882
56	1 11 51 36	1 117 583 616	.4961536	.1832868	21	1 25 66 41	1 408 694 561	.4813381	.3880781
57	1 11 72 49	1 180 932 193	.5115364	.1865002	22	1 25 88 84	1 412 467 848	.4962684	.3911661
58	1 11 93 64	1 184 287 112	.5269119	.1897116	23	1 26 11 29	1 416 247 867	.5111921	.3942523
59	1 12 14 81	1 187 648 379	.5422802	.1929209	24	1 26 33 76	1 420 034 624	.5261092	.3973366
1060	1 12 36 00	1 191 016 000	32.5576412	10.1961283	1125	1 26 56 25	1 423 828 125	33.5410196	10.4004192
61	1 12 57 21	1 194 389 981	.5729949	.1993336	26	1 26 78 76	1 427 628 376	.5559234	.4034999
62	1 12 78 44	1 197 770 328	.5883415	.2025369	27	1 27 01 29	1 431 435 383	.5708206	.4065787
63	1 12 99 69	1 201 157 047	.6036807	.2057382	28	1 27 23 84	1 435 249 152	.5857112	.4096557
64	1 13 20 96	1 204 550 144	.6190129	.2089375	29	1 27 46 41	1 439 069 689	.6005952	.4127310
1065	1 13 42 25	1 207 949 625	32.6343377	10.2121347	1130	1 27 69 00	1 442 897 000	33.6154726	10.4158044
66	1 13 63 56	1 211 355 496	.6496554	.2153300	31	1 27 91 61	1 446 731 091	.6303434	.4188760
67	1 13 84 89	1 214 767 763	.6649659	.2185233	32	1 28 14 24	1 450 571 968	.6452077	.4219458
68	1 14 06 24	1 218 186 432	.6802693	.2217146	33	1 28 36 89	1 454 419 637	.6600653	.4250138
69	1 14 27 61	1 221 611 509	.6955654	.2249039	34	1 28 59 56	1 458 274 104	.6749165	.4280800
1070	1 14 49 00	1 225 043 000	32.7108544	10.2280912	1135	1 28 82 25	1 462 135 375	33.6897610	10.4311443
71	1 14 70 41	1 228 480 911	.7261363	.2312766	36	1 29 04 96	1 466 003 456	.7045991	.4342069
72	1 14 91 84	1 231 925 248	.7414111	.2344599	37	1 29 27 69	1 469 878 353	.7194306	.4372677
73	1 15 13 29	1 235 376 017	.7566787	.2376413	38	1 29 50 44	1 473 760 072	.7342556	.4403267
74	1 15 34 76	1 238 833 224	.7719392	.2408207	39	1 29 73 21	1 477 648 619	.7490741	.4433839
1075	1 15 56 25	1 242 296 875	32.7871926	10.2439981	1140	1 29 96 00	1 481 544 000	33.7638860	10.4464393
76	1 15 77 76	1 245 766 976	.8024389	.2471735	41	1 30 18 81	1 485 446 221	.7786915	.4494929
77	1 15 99 29	1 249 243 533	.8176782	.2503470	42	1 30 41 64	1 489 355 288	7934905	4525448
78	1 16 20 84	1 252 726 552	.8329103	.2535186	43	1 30 64 49	1 493 271 207	.8082830	.4555948
79	1 16 42 41	1 256 216 039	.8481354	.2566881	44	1 30 87 36	1 497 193 984	.8230691	.4586431
1080	1 16 64 00	1 259 712 000	32.8633535	10.2598557	1145	1 31 10 25	1 501 123 625	33.8378486	10.4616896
81	1 16 85 61	1 263 214 441	.8785644	.2630213	46	1 31 33 16	1 505 060 136	.8526218	.4647343
82	1 17 07 24	1 266 723 368	.8937684	.2661850	47	1 31 56 09	1 509 003 523	.8673884	.4677773
83	1 17 28 89	1 270 238 787	.9089653	.2693467	48	1 31 79 04	1 512 953 792	.8821487	.4708185
84	1 17 50 56	1 273 760 704	.9241553	.2725065	49	1 32 02 01	1 516 910 949	.8969025	.4738579
1085	1 17 72 25	1 277 289 125	32.9393382	10.2756644	1150	1 32 25 00	1 520 875 000	33.9116499	10.4768955
86	1 17 93 96	1 280 824 056	.9545141	.2788203	51	1 32 48 01	1 524 845 951	.9263909	.4799314
87	1 18 15 69	1 284 365 503	.9696830	.2819743	52	1 32 71 04	1 528 823 808	.9411255	.4829656
88	1 18 37 44	1 287 913 472	.9848450	.2851264	53	1 32 94 09	1 532 808 577	.9558537	.4859980
89	1 18 59 21	1 291 467 969	33.0000000	.2882765	54	1 33 17 16	1 536 800 264	.9705755	.4890286
1090	1 18 81 00	1 295 029 000	33.0151480	10.2914247	1155	1 33 40 25	1 540 798 875	33.9852910	10.4920575
91	1 19 02 81	1 298 596 571	.0302891	.2945709	56	1 33 63 36	1 544 804 416	34.0000000	.4950847
92	1 19 24 64	1 302 170 688	.0454233	.2977153	57	1 33 86 49	1 548 816 893	.0147027	.4981101
93	1 19 46 49	1 305 751 357	.0605505	.3008577	58	1 34 09 64	1 552 836 312	.0293990	.5011337
94	1 19 68 36	1 309 338 584	.0756708	.3039982	59	1 34 32 81	1 556 862 679	.0440890	.5041556
1095	1 19 90 25	1 312 932 375	33.0907842	10.3071368	1160	1 34 56 00	1 560 896 000	34.0587727	10.5071757
96	1 20 12 16	1 316 532 736	.1058907	.3102735	61	1 34 79 21	1 564 936 281	.0734501	.5101942
97	1 20 34 09	1 320 139 673	.1209903	.3134083	62	1 35 02 44	1 568 983 528	.0881211	.5132109
98	1 20 56 04	1 323 753 192	.1360830	.3165411	63	1 35 25 69	1 573 037 747	.1027858	.5162259
99	1 20 78 01	1 327 373 299	.1511689	.3196721	64	1 35 48 96	1 577 098 944	.1174442	.5192391
1100	1 21 00 00	1 331 000 000	33.1662479	10.3228012	1165	1 35 72 25	1 581 167 125	34.1320963	10.5222506
1	1 21 42 01	1 334 633 301	.1813200	.3259284	66	1 35 95 56	1 585 242 296	.1467422	.5252604
2	1 21 44 04	1 338 273 208	.1963853	.3290537	67	1 36 18 89	1 589 324 463	.1613817	.5282685
3	1 21 66 09	1 341 919 727	.2114438	.3321770	68	1 36 42 24	1 593 413 632	.1760150	.5312749
4	1 21 88 16	1 345 572 864	.2264955	.3352985	69	1 36 65 61	1 597 509 809	.1906420	.5342795
1105	1 22 10 25	1 349 232 625	33.2415403	10.3384181	1170	1 36 89 00	1 601 613 000	34.2052627	10.5372825

TABLE III

SQUARES, CUBES, SQUARE ROOTS, CUBE ROOTS, OF NUMBERS—*Continued*

No.	Square	Cube.	Sq. Rt.	Cu. Rt.	No.	Square	Cube.	Sq. Rt.	Cu. Rt.
1170	1 36 89 00	1 601 613 000	34.2052627	10.5372825	1235	1 52 52 25	1 883 652 875	35.1425668	10.7289112
71	1 37 12 41	1 605 723 211	.2198773	.5402837	36	1 52 76 96	1 888 232 256	.1567917	.7318062
72	1 37 35 84	1 609 840 448	.2344855	.5432832	37	1 53 01 69	1 892 819 053	.1710108	.7346997
73	1 37 59 29	1 613 964 717	.2490875	.5462810	38	1 53 26 44	1 897 413 272	.1852242	.7375916
74	1 37 82 76	1 618 096 024	.2636834	.5492771	39	1 53 51 21	1 902 014 919	.1994318	.7404819
1175	1 38 06 25	1 622 234 375	34.2782730	10.5522715	1240	1 53 76 00	1 906 624 000	35.2136337	10.7433707
76	1 38 29 76	1 626 379 776	.2928564	.5552642	41	1 54 00 81	1 911 240 521	.2278299	.7462579
77	1 38 53 29	1 630 532 233	.3074336	.5582552	42	1 54 25 64	1 915 864 488	.2420204	.7491436
78	1 38 76 84	1 634 691 752	.3220046	.5612445	43	1 54 50 49	1 920 495 907	.2562051	.7520277
79	1 39 00 41	1 638 858 339	.3365694	.5642322	44	1 54 75 36	1 925 134 784	.2703842	.7549103
1180	1 39 24 00	1 643 032 000	34.3511281	10.5672181	1245	1 55 00 25	1 929 781 125	35.2845575	10.7577913
81	1 39 47 61	1 647 212 741	.3656805	.5702024	46	1 55 25 16	1 934 434 936	.2987252	.7606708
82	1 39 71 24	1 651 400 568	.3802268	.5731849	47	1 55 50 09	1 939 096 223	.3128872	.7635488
83	1 39 94 89	1 655 595 487	.3947670	.5761658	48	1 55 75 04	1 943 764 992	.3270435	.7664252
84	1 40 18 56	1 659 797 504	.4093011	.5791449	49	1 56 00 01	1 948 441 249	.3411941	.7693001
1185	1 40 42 25	1 664 006 625	34.4238289	10.5821225	1250	1 56 25 00	1 953 125 000	35.3553391	10.7721735
86	1 40 65 96	1 668 222 856	.4383507	.5850983	51	1 56 50 01	1 957 816 251	.3694784	.7750453
87	1 40 89 69	1 672 446 203	.4528663	.5880725	52	1 56 75 04	1 962 515 008	.3836120	.7779156
88	1 41 13 44	1 676 676 672	.4673759	.5910450	53	1 57 00 09	1 967 221 277	.3977400	.7807843
89	1 41 37 21	1 680 914 269	.4818793	.5940158	54	1 57 25 16	1 971 935 064	.4118624	.7836516
1190	1 41 61 00	1 685 159 000	34.4963766	10.5969850	1255	1 57 50 25	1 976 656 375	35.4259792	10.7865173
91	1 41 84 81	1 689 410 871	.5108678	.5999525	56	1 57 75 36	1 981 385 216	.4400903	.7893815
92	1 42 08 64	1 693 669 888	.5253530	.6029184	57	1 58 00 49	1 986 121 593	.4541958	.7922441
93	1 42 32 49	1 697 936 057	.5398321	.6058826	58	1 58 25 64	1 990 865 512	.4682957	.7951053
94	1 42 56 36	1 702 209 384	.5543051	.6088451	59	1 58 50 81	1 995 616 979	.4823900	.7979649
1195	1 42 80 25	1 706 489 875	34.5687720	10.6118060	1260	1 58 76 00	2 000 376 000	35.4964787	10.8008230
96	1 43 04 16	1 710 777 536	.5832329	.6147652	61	1 59 01 21	2 005 142 581	.5105618	.8036797
97	1 43 28 09	1 715 072 373	.5976879	.6177228	62	1 59 26 44	2 009 916 728	.5246393	.8065348
98	1 43 52 04	1 719 374 392	.6121366	.6206788	63	1 59 51 69	2 014 698 447	.5387113	.8093884
99	1 43 76 01	1 723 683 599	.6265794	.6236331	64	1 59 76 96	2 019 487 744	.5527777	.8122404
1200	1 44 00 00	1 728 000 000	34.6410162	10.6265857	1265	1 60 02 25	2 024 284 625	35.5668385	10.8150909
1	1 44 24 01	1 732 323 601	.6554469	.6295367	66	1 60 27 56	2 029 089 096	.5808937	.8179400
2	1 44 48 04	1 736 654 408	.6698716	.6324860	67	1 60 52 89	2 033 901 163	.5949434	.8207876
3	1 44 72 09	1 740 992 427	.6842904	.6354338	68	1 60 78 24	2 038 720 832	.6089876	.8236336
4	1 44 96 16	1 745 337 664	.6987031	.6383799	69	1 61 03 61	2 043 548 109	.6230262	.8264782
1205	1 45 20 25	1 749 690 125	34.7131099	10.6413244	1270	1 61 29 00	2 048 383 000	35.6370593	10.8293213
6	1 45 44 36	1 754 049 816	.7275107	.6442672	71	1 61 54 41	2 053 225 511	.6510869	.8321629
7	1 45 68 49	1 758 416 743	.7419055	.6472085	72	1 61 79 84	2 058 075 648	.6651090	.8350030
8	1 45 92 64	1 762 790 912	.7562944	.6501480	73	1 62 05 29	2 062 933 417	.6791255	.8378416
9	1 46 16 81	1 767 172 329	.7706773	.6530860	74	1 62 30 76	2 067 798 824	.6931366	.8406788
1210	1 46 41 00	1 771 561 000	34.7850543	10.6560223	1275	1 62 56 25	2 072 671 875	35.7071421	10.8435144
11	1 46 65 21	1 775 956 931	.7994253	.6589570	76	1 62 81 76	2 077 552 576	.7211422	.8463485
12	1 46 89 44	1 780 360 128	.8137904	.6618902	77	1 63 07 29	2 082 440 933	.7351367	.8491812
13	1 47 13 69	1 784 770 597	.8281495	.6648217	78	1 63 32 84	2 087 336 952	.7491258	.8520125
14	1 47 37 96	1 789 188 344	.8425028	.6677516	79	1 63 58 41	2 092 240 639	.7631095	.8548422
1215	1 47 62 25	1 793 613 375	34.8568501	10.6706799	1280	1 63 84 00	2 097 152 000	35.7770876	10.8576704
16	1 47 86 56	1 798 045 696	.8711915	.6736066	81	1 64 09 61	2 102 071 041	.7910603	.8604972
17	1 48 10 89	1 802 485 313	.8855271	.6765317	82	1 64 35 24	2 106 997 768	.8050276	.8633225
18	1 48 35 24	1 806 932 232	.8998567	.6794552	83	1 64 60 89	2 111 932 187	.8189894	.8661464
19	1 48 59 61	1 811 386 459	.9141805	.6823771	84	1 64 86 56	2 116 874 304	.8329457	.8689687
1220	1 48 84 00	1 815 848 000	34.9284984	10.6852973	1285	1 65 12 25	2 121 824 125	35.8468966	10.8717897
21	1 49 08 41	1 820 316 861	.9428104	.6882160	86	1 65 37 96	2 126 781 656	.8608421	.8746091
22	1 49 32 84	1 824 793 048	.9571166	.6911331	87	1 65 63 69	2 131 746 903	.8747822	.8774271
23	1 49 57 29	1 829 276 567	.9714169	.6940486	88	1 65 89 44	2 136 719 872	.8887169	.8802436
24	1 49 81 76	1 833 767 424	.9857114	.6969625	89	1 66 15 21	2 141 700 569	.9026461	.8830587
1225	1 50 06 25	1 838 265 625	35.0000000	10.6998748	1290	1 66 41 00	2 146 689 000	35.9165699	10.8858723
26	1 50 30 76	1 842 771 176	.0142828	.7027855	91	1 66 66 81	2 151 685 171	.9304884	.8886845
27	1 50 55 29	1 847 284 083	.0285598	.7056947	92	1 66 92 64	2 156 689 088	.9444015	.8914952
28	1 50 79 84	1 851 804 352	.0428309	.7086023	93	1 67 18 49	2 161 700 757	.9583092	.8943044
29	1 51 04 41	1 856 331 989	.0570963	.7115083	94	1 67 44 36	2 166 720 184	.9722115	.8971123
1230	1 51 29 00	1 860 867 000	35.0713558	10.7144127	1295	1 67 70 25	2 171 747 375	35.9861084	10.8999186
31	1 51 53 61	1 865 409 391	.0856096	.7173155	96	1 67 96 16	2 176 782 336	36.0000000	.9027235
32	1 51 78 24	1 869 959 168	.0998575	.7202168	97	1 68 22 09	2 181 825 073	.0138862	.9055269
33	1 52 02 89	1 874 516 337	.1140997	.7231165	98	1 68 48 04	2 186 875 592	.0277671	.9083290
34	1 52 27 56	1 879 080 904	.1283361	.7260146	99	1 68 74 01	2 191 933 899	.0416426	.9111296
1235	1 52 52 25	1 883 652 875	35.1425668	10.7289112	1300	1 69 00 00	2 197 000 000	36.0555128	10.9139287

TABLE III

SQUARES, CUBES, SQUARE ROOTS, CUBE ROOTS, OF NUMBERS—*Continued*

No.	Square	Cube.	Sq. Rt.	Cu. Rt.	No.	Square	Cube.	Sq. Rt.	Cu. Rt.
1300	1 69 00 00	2 197 000 000	36.0555128	10.9139287	1365	1 86 32 25	2 543 302 125	36.9459064	11.0928775
1	1 69 26 01	2 202 073 901	.0693776	.9167265	66	1 86 59 56	2 548 895 896	.9594372	.0955857
2	1 69 52 04	2 207 155 608	.0832371	.9195228	67	1 86 86 89	2 554 497 863	.9729631	.0982926
3	1 69 78 09	2 212 245 127	.0970913	.9223177	68	1 87 14 24	2 560 108 032	.9864840	.1009982
4	1 70 04 16	2 217 342 464	.1109402	.9251111	69	1 87 41 61	2 565 726 409	37.0000000	.1037025
1305	1 70 30 25	2 222 447 625	36.1247837	10.9279031	1370	1 87 69 00	2 571 353 000	37.0135110	11.1064054
6	1 70 56 36	2 227 560 616	.1386220	.9306937	71	1 87 96 41	2 576 987 811	.0270172	.1091070
7	1 70 82 49	2 232 681 443	.1524550	.9334829	72	1 88 23 84	2 582 630 848	.0405184	.1118073
8	1 71 08 64	2 237 810 112	.1662826	.9362706	73	1 88 51 29	2 588 282 117	.0540146	.1145064
9	1 71 34 81	2 242 946 629	.1801050	.9390569	74	1 88 78 76	2 593 941 624	.0675060	.1172041
1310	1 71 61 00	2 248 091 000	36.1939221	10.9418418	1375	1 89 06 25	2 599 609 375	37.0809924	11.1199004
11	1 71 87 21	2 253 243 231	.2077340	.9446253	76	1 89 33 76	2 605 285 376	.0944740	.1225955
12	1 72 13 44	2 258 403 328	.2215406	.9474074	77	1 89 61 29	2 610 969 633	.1079506	.1252893
13	1 72 39 69	2 263 571 297	.2353419	.9501880	78	1 89 88 84	2 616 662 152	.1214224	.1279817
14	1 72 65 96	2 268 747 144	.2491379	.9529673	79	1 90 16 41	2 622 362 939	.1348893	.1306729
1315	1 72 92 25	2 273 930 875	36.2629287	10.9557451	1380	1 90 44 00	2 628 072 000	37.1483512	11.1333628
16	1 73 18 56	2 279 122 496	.2767143	.9585215	81	1 90 71 61	2 633 789 341	.1618084	.1360514
17	1 73 44 89	2 284 322 013	.2904946	.9612965	82	1 90 99 24	2 639 514 968	.1752606	.1387386
18	1 73 71 24	2 289 529 432	.3042697	.9640701	83	1 91 26 89	2 645 248 887	.1887079	.1414246
19	1 73 97 61	2 294 744 759	.3180396	.9668423	84	1 91 54 56	2 650 991 104	.2021505	.1441093
1320	1 74 24 00	2 299 968 000	36.3318042	10.9696131	1385	1 91 82 25	2 656 741 625	37.2155881	11.1467926
21	1 74 50 41	2 305 199 161	.3455637	.9723825	86	1 92 09 96	2 662 500 456	.2290209	.1494747
22	1 74 76 84	2 310 438 248	.3593179	.9751505	87	1 92 37 69	2 668 267 603	.2424489	.1521555
23	1 75 03 29	2 315 685 267	.3730670	.9779171	88	1 92 65 44	2 674 043 072	.2558720	.1548350
24	1 75 29 76	2 320 940 224	.3868108	.9806823	89	1 92 93 21	2 679 826 869	.2692903	.1575133
1325	1 75 56 25	2 326 203 125	36.4005494	10.9834462	1390	1 93 21 00	2 685 619 000	37.2827037	11.1601903
26	1 75 82 76	2 331 473 976	.4142829	.9862086	91	1 93 48 81	2 691 419 471	.2961124	.1628659
27	1 76 09 29	2 336 752 783	.4280112	.9889696	92	1 93 76 64	2 697 228 288	.3095162	.1655403
28	1 76 35 84	2 342 039 552	.4417343	.9917293	93	1 94 04 49	2 703 045 457	.3229152	.1682134
29	1 76 62 41	2 347 334 289	.4554523	.9944876	94	1 94 32 36	2 708 870 984	.3363094	.1708852
1330	1 76 89 00	2 352 637 000	36.4691650	10.9972445	1395	1 94 60 25	2 714 704 875	37.3496988	11.1735558
31	1 77 15 61	2 357 947 691	.4828727	11.0000000	96	1 94 88 16	2 720 547 136	.3630834	.1762250
32	1 77 42 24	2 363 266 368	.4965752	.0027541	97	1 95 16 09	2 726 397 773	.3764632	.1788930
33	1 77 68 89	2 368 593 037	.5102725	.0055069	98	1 95 44 04	2 732 256 792	.3898382	.1815598
34	1 77 95 56	2 373 927 704	.5239647	.0082583	99	1 95 72 01	2 738 124 199	.4032084	.1842252
1335	1 78 22 25	2 379 270 375	36.5376518	11.0110082	1400	1 96 00 00	2 744 000 000	37.4165738	11.1868894
36	1 78 48 96	2 384 621 056	.5513338	.0137569	1	1 96 28 01	2 749 884 201	.4299345	.1895523
37	1 78 75 69	2 389 979 753	.5650106	.0165041	2	1 96 56 04	2 755 776 808	.4432904	.1922139
38	1 79 02 44	2 395 346 472	.5786823	.0192500	3	1 96 84 09	2 761 677 827	.4566416	.1948743
39	1 79 29 21	2 400 721 219	.5923489	.0219945	4	1 97 12 16	2 767 587 264	.4699880	.1975334
1340	1 79 56 00	2 406 104 000	36.6060104	11.0247377	1405	1 97 40 25	2 773 505 125	37.4833296	11.2001913
41	1 79 82 81	2 411 494 821	.6196668	.0274795	6	1 97 68 36	2 779 431 416	.4966665	.2028479
42	1 80 09 64	2 416 893 688	.6333181	.0302199	7	1 97 96 49	2 785 366 143	.5099987	.2055032
43	1 80 36 49	2 422 300 607	.6469644	.0329590	8	1 98 24 64	2 791 309 312	.5233261	.2081573
44	1 80 63 36	2 427 715 584	.6606056	.0356967	9	1 98 52 81	2 797 260 929	.5366487	.2108101
1345	1 80 90 25	2 433 138 625	36.6742416	11.0384330	1410	1 98 81 00	2 803 221 000	37.5499667	11.2134617
46	1 81 17 16	2 438 569 736	.6878726	.0411680	11	1 99 09 21	2 809 189 531	.5632799	.2161120
47	1 81 44 09	2 444 008 923	.7014986	.0439017	12	1 99 37 44	2 815 166 528	.5765885	.2187611
48	1 81 71 04	2 449 456 192	.7151195	.0466339	13	1 99 65 69	2 821 151 997	.5898922	.2214089
49	1 81 98 01	2 454 911 549	.7287353	.0493649	14	1 99 93 96	2 827 145 944	.6031913	.2240554
1350	1 82 25 00	2 460 375 000	36.7423461	11.0520945	1415	2 00 22 25	2 833 148 375	37.6164857	11.2267007
51	1 82 52 01	2 465 846 551	.7559519	.0548227	16	2 00 50 56	2 839 159 296	.6297754	.2293448
52	1 82 79 04	2 471 326 208	.7695526	.0575497	17	2 00 78 89	2 845 178 713	.6430604	.2319876
53	1 83 06 09	2 476 813 977	.7831483	.0602752	18	2 01 07 24	2 851 206 632	.6563407	.2346292
54	1 83 33 16	2 482 309 864	.7967390	.0629994	19	2 01 35 61	2 857 243 059	.6696164	.2372696
1355	1 83 60 25	2 487 813 875	36.8103246	11.0657222	1420	2 01 64 00	2 863 288 000	37.6828874	11.2399087
56	1 83 87 36	2 493 326 016	.8239053	.0684437	21	2 01 92 41	2 869 341 461	.6961536	.2425465
57	1 84 14 49	2 498 846 293	.8374809	.0711639	22	2 02 20 84	2 875 403 448	.7094153	.2451831
58	1 84 41 64	2 504 374 712	.8510515	.0738828	23	2 02 49 29	2 881 473 967	.7226722	.2478185
59	1 84 68 81	2 509 911 279	.8646172	.0766003	24	2 02 77 76	2 887 553 024	.7359245	.2504527
1360	1 84 96 00	2 515 456 000	36.8781778	11.0793165	1425	2 03 06 25	2 893 640 625	37.7491722	11.2530856
61	1 85 23 21	2 521 008 881	.8917335	.0820314	26	2 03 34 76	2 899 736 776	.7624152	.2557173
62	1 85 50 44	2 526 569 928	.9052842	.0847449	27	2 03 63 29	2 905 841 483	.7756535	.2583478
63	1 85 77 69	2 532 139 147	.9188299	.0874571	28	2 03 91 84	2 911 954 752	.7888873	.2609770
64	1 86 04 96	2 537 716 544	.9323706	.0901679	29	2 04 20 41	2 918 076 589	.8021163	.2636050
1365	1 86 32 25	2 543 302 125	36.9459064	11.0928775	1430	2 04 49 00	2 924 207 000	37.8153408	11.2662318

TABLE III

SQUARES, CUBES, SQUARE ROOTS, CUBE ROOTS, OF NUMBERS—*Concluded*

No.	Square	Cube.	Sq. Rt.	Cu. Rt.	No.	Square	Cube.	Sq. Rt.	Cu. Rt.
1430	2 04 49 00	2 924 207 000	37.8153408	11.2662318	1495	2 23 50 25	3 341 362 375	38.6652299	11.4344092
31	2 04 77 61	2 930 345 991	.8285606	.2688573	96	2 23 80 16	3 348 071 936	.6781593	.4369581
32	2 05 06 24	2 936 493 568	.8417759	.2714816	97	2 24 10 09	3 354 790 473	.6910843	.4395059
33	2 05 34 89	2 942 649 737	.8549864	.2741047	98	2 24 40 04	3 361 517 992	.7040050	.4420525
34	2 05 63 56	2 948 814 504	.8681924	.2767266	99	2 24 70 01	3 368 254 499	.7169214	.4445980
1435	2 05 92 25	2 954 987 875	37.8813938	11.2793472	1500	2 25 00 00	3 375 000 000	38.7298335	11.4471424
36	2 06 20 96	2 961 169 856	.8945906	.2819666	1	2 25 30 01	3 381 754 501	.7427412	.4496857
37	2 06 49 69	2 967 360 453	.9077828	.2845849	2	2 25 60 04	3 388 518 008	.7556447	.4522278
38	2 06 78 44	2 973 559 672	.9209704	.2872019	3	2 25 90 09	3 395 290 527	.7685439	.4547688
39	2 07 07 21	2 979 767 519	.9341535	.2898177	4	2 26 20 16	3 402 072 064	.7814389	.4573087
1440	2 07 36 00	2 985 984 000	37.9473319	11.2924323	1505	2 26 50 25	3 408 862 625	38.7943294	11.4598474
41	2 07 64 81	2 992 209 121	.9605058	.2950457	6	2 26 80 36	3 415 662 216	.8072158	.4623850
42	2 07 93 64	2 998 442 888	.9736751	.2976579	7	2 27 10 49	3 422 470 843	.8200978	.4649215
43	2 08 22 49	3 004 685 307	.9868398	.3002688	8	2 27 40 64	3 429 288 512	.8329757	.4674568
44	2 08 51 36	3 010 936 384	38.0000000	.3028786	9	2 27 70 81	3 436 115 229	.8458491	.4699911
1445	2 08 80 25	3 017 196 125	38.0131556	11.3054871	1510	2 28 01 00	3 442 951 000	38.8587184	11.4725242
46	2 09 09 16	3 023 464 536	.0263067	.3080945	11	2 28 31 21	3 449 795 831	.8715834	.4750562
47	2 09 38 09	3 029 741 623	.0394532	.3107006	12	2 28 61 44	3 456 649 728	.8844442	.4775871
48	2 09 67 04	3 036 027 392	.0525952	.3133056	13	2 28 91 69	3 463 512 697	.8973006	.4801169
49	2 09 96 01	3 042 321 849	.0657326	.3159094	14	2 29 21 96	3 470 384 744	.9101529	.4826455
1450	2 10 25 00	3 048 625 000	38.0788655	11.3185119	1515	2 29 52 25	3 477 265 875	38.9230009	11.4851731
51	2 10 54 01	3 054 936 851	.0919939	.3211132	16	2 29 82 56	3 484 156 096	.9358447	.4876995
52	2 10 83 04	3 061 257 408	.1051178	.3237134	17	2 30 12 89	3 491 055 413	.9486841	.4902249
53	2 11 12 09	3 067 586 677	.1182371	.3263124	18	2 30 43 24	3 497 963 832	.9615194	.4927491
54	2 11 41 16	3 073 924 664	.1313519	.3289102	19	2 30 73 61	3 504 881 359	.9743505	.4952722
1455	2 11 70 25	3 080 271 375	38.1444622	11.3315067	1520	2 31 04 00	3 511 808 000	38.9871774	11.4977942
56	2 11 99 36	3 086 626 816	.1575681	.3341022	21	2 31 34 41	3 518 743 761	39.0000000	.5003151
57	2 12 28 49	3 092 990 993	.1706693	.3366964	22	2 31 64 84	3 525 688 648	.0128184	.5028348
58	2 12 57 64	3 099 363 912	.1837662	.3392894	23	2 31 95 29	3 532 642 667	.0256326	.5053535
59	2 12 86 81	3 105 745 579	.1968585	.3418813	24	2 32 25 76	3 539 605 824	.0384426	.5078711
1460	2 13 16 00	3 112 136 000	38.2099463	11.3444719	1525	2 32 56 25	3 546 578 125	39.0512483	11.5103876
61	2 13 45 21	3 118 535 181	.2230297	.3470614	26	2 32 86 76	3 553 559 576	.0640499	.5129030
62	2 13 74 44	3 124 943 128	.2361085	.3496497	27	2 33 17 29	3 560 550 183	.0768473	.5154173
63	2 14 03 69	3 131 359 847	.2491829	.3522368	28	2 33 47 84	3 567 549 952	.0896406	.5179305
64	2 14 32 96	3 137 785 344	.2622529	.3548227	29	2 33 78 41	3 574 558 889	.1024296	.5204425
1465	2 14 62 25	3 144 219 625	38.2753184	11.3574075	1530	2 34 09 00	3 581 577 000	39.1152144	11.5229535
66	2 14 91 56	3 150 662 696	.2883794	.3599911	31	2 34 39 61	3 588 604 291	.1279951	.5254634
67	2 15 20 89	3 157 114 563	.3014360	.3625735	32	2 34 70 24	3 595 640 768	.1407716	.5279722
68	2 15 50 24	3 163 575 232	.3144881	.3651547	33	2 35 00 89	3 602 686 437	.1535439	.5304799
69	2 15 79 61	3 170 044 709	.3275358	.3677347	34	2 35 31 56	3 609 741 304	.1663120	.5329865
1470	2 16 09 00	3 176 523 000	38.3405790	11.3703136	1535	2 35 62 25	3 616 805 375	39.1790760	11.5354920
71	2 16 38 41	3 183 010 111	.3536178	.3728914	36	2 35 92 96	3 623 878 656	.1918359	.5379965
72	2 16 67 84	3 189 506 048	.3666522	.3754679	37	2 36 23 69	3 630 961 153	.2045915	.5404998
73	2 16 97 29	3 196 010 817	.3796821	.3780433	38	2 36 54 44	3 638 052 872	.2173431	.5430021
74	2 17 26 76	3 202 524 424	.3927076	.3806175	39	2 36 85 21	3 645 153 819	.2300905	.5455033
1475	2 17 56 25	3 209 046 875	38.4057287	11.3831906	1540	2 37 16 00	3 652 264 000	39.2428337	11.5480034
76	2 17 85 76	3 215 578 176	.4187454	.3857625	41	2 37 46 81	3 659 383 421	.2555728	.5505025
77	2 18 15 29	3 222 118 333	.4317577	.3883332	42	2 37 77 64	3 666 512 088	.2683078	.5530004
78	2 18 44 84	3 228 667 352	.4447656	.3909028	43	2 38 08 49	3 673 650 007	.2810387	.5554973
79	2 18 74 41	3 235 225 239	.4577691	.3934712	44	2 38 39 36	3 680 797 184	.2937654	.5579931
1480	2 19 04 00	3 241 792 000	38.4707681	11.3960384	1545	2 38 70 25	3 687 953 625	39.3064880	11.5604878
81	2 19 33 61	3 248 367 641	.4837627	.3986045	46	2 39 01 16	3 695 119 336	.3192065	.5629815
82	2 19 63 24	3 254 952 168	.4967530	.4011695	47	2 39 32 09	3 702 294 323	.3319208	.5654740
83	2 19 92 89	3 261 545 587	.5097390	.4037332	48	2 39 63 04	3 709 478 592	.3446311	.5679655
84	2 20 22 56	3 268 147 904	.5227206	.4062959	49	2 39 94 01	3 716 672 149	.3573373	.5704559
1485	2 20 52 25	3 274 759 125	38.5356977	11.4088574	1550	2 40 25 00	3 723 875 000	39.3700394	11.5729453
86	2 20 81 96	3 281 379 256	.5486705	.4114177	51	2 40 56 01	3 731 087 151	.3827373	.5754336
87	2 21 11 69	3 288 008 303	.5616389	.4139769	52	2 40 87 04	3 738 308 608	.3954312	.5779208
88	2 21 41 44	3 294 646 272	.5746030	.4165349	53	2 41 18 09	3 745 539 377	.4081210	.5804069
89	2 21 71 21	3 301 293 169	.5875627	.4190918	54	2 41 49 16	3 752 779 464	.4208067	.5828919
1490	2 22 01 00	3 307 949 000	38.6005181	11.4216476	1555	2 41 80 25	3 760 028 875	39.4334883	11.5853759
91	2 22 30 81	3 314 613 771	.6134691	.4242022	56	2 42 11 36	3 767 287 616	.4461658	.5878588
92	2 22 60 64	3 321 287 488	.6264158	.4267556	57	2 42 42 49	3 774 555 693	.4588393	.5903407
93	2 22 90 49	3 327 970 157	.6393582	.4293079	58	2 42 73 64	3 781 833 112	.4715087	.5928215
94	2 23 20 36	3 334 661 784	.6522962	.4318591	59	2 43 04 81	3 789 119 879	.4841740	.5953013
1495	2 23 50 25	3 341 362 375	38.6652299	11.4344092	1560	2 43 36 00	3 796 416 000	39.4968353	11.5977799

TABLE IV

LOGARITHMS 100 TO 1000

	0	1	2	3	4	5	6	7	8	9	1	2	3	4	5	6	7	8	9
10	0000	0043	0086	0128	0170	0212	0253	0294	0334	0374									
11	0414	0453	0492	0531	0569	0607	0645	0682	0719	0755	4	8	11	15	19	23	26	30	34
12	0792	0828	0864	0899	0934	0969	1004	1038	1072	1106	3	7	10	14	17	21	24	28	31
13	1139	1173	1206	1239	1271	1303	1335	1367	1399	1430	3	6	10	13	16	19	23	26	29
14	1461	1492	1523	1553	1584	1614	1644	1673	1703	1732	3	6	9	12	15	18	21	24	27
15	1761	1790	1818	1847	1875	1903	1931	1959	1987	2014	3	6	8	11	14	17	20	22	25
16	2041	2068	2095	2122	2148	2175	2201	2227	2253	2279	3	5	8	11	13	16	18	21	24
17	2304	2330	2355	2380	2405	2430	2455	2480	2504	2529	2	5	7	10	12	15	17	20	22
18	2553	2577	2601	2625	2648	2672	2695	2718	2742	2765	2	5	7	9	12	14	16	19	21
19	2788	2810	2833	2856	2878	2900	2923	2945	2967	2989	2	4	7	9	11	13	16	18	20
20	3010	3032	3054	3075	3096	3118	3139	3160	3181	3201	2	4	6	8	11	13	15	17	19
21	3222	3243	3263	3284	3304	3324	3345	3365	3385	3404	2	4	6	8	10	12	14	16	18
22	3424	3444	3464	3483	3502	3522	3541	3560	3579	3598	2	4	6	8	10	12	14	15	17
23	3617	3636	3655	3674	3692	3711	3729	3747	3766	3784	2	4	6	7	9	11	13	15	17
24	3802	3820	3838	3856	3874	3892	3909	3927	3945	3962	2	4	5	7	9	11	12	14	16
25	3979	3997	4014	4031	4048	4065	4082	4099	4116	4133	2	3	5	7	9	10	12	14	15
26	4150	4166	4183	4200	4216	4232	4249	4265	4281	4298	2	3	5	7	8	10	11	13	15
27	4314	4330	4346	4362	4378	4393	4409	4425	4440	4456	2	3	5	6	8	9	11	13	14
28	4472	4487	4502	4518	4533	4548	4564	4579	4594	4609	2	3	5	6	8	9	11	12	14
29	4624	4639	4654	4669	4683	4698	4713	4728	4742	4757	1	3	4	6	7	9	10	12	13
30	4771	4786	4800	4814	4829	4843	4857	4871	4886	4900	1	3	4	6	7	9	10	11	13
31	4914	4928	4942	4955	4969	4983	4997	5011	5024	5038	1	3	4	6	7	8	10	11	12
32	5051	5065	5079	5092	5105	5119	5132	5145	5159	5172	1	3	4	5	7	8	9	11	12
33	5185	5198	5211	5224	5237	5250	5263	5276	5289	5302	1	3	4	5	6	8	9	10	12
34	5315	5328	5340	5353	5366	5378	5391	5403	5416	5428	1	3	4	5	6	8	9	10	11
35	5441	5453	5465	5478	5490	5502	5514	5527	5539	5551	1	2	4	5	6	7	9	10	11
36	5563	5575	5587	5599	5611	5623	5635	5647	5658	5670	1	2	4	5	6	7	8	10	11
37	5682	5694	5705	5717	5729	5740	5752	5763	5775	5786	1	2	3	5	6	7	8	9	10
38	5798	5809	5821	5832	5843	5855	5866	5877	5888	5899	1	2	3	5	6	7	8	9	10
39	5911	5922	5933	5944	5955	5966	5977	5988	5999	6010	1	2	3	4	5	7	8	9	10
40	6021	6031	6042	6053	6064	6075	6085	6096	6107	6117	1	2	3	4	5	6	8	9	10
41	6128	6138	6149	6160	6170	6180	6191	6201	6212	6222	1	2	3	4	5	6	7	8	9
42	6232	6243	6253	6263	6274	6284	6294	6304	6314	6325	1	2	3	4	5	6	7	8	9
43	6335	6345	6355	6365	6375	6385	6395	6405	6415	6425	1	2	3	4	5	6	7	8	9
44	6435	6444	6454	6464	6474	6484	6493	6503	6513	6522	1	2	3	4	5	6	7	8	9
45	6532	6542	6551	6561	6571	6580	6590	6599	6609	6618	1	2	3	4	5	6	7	8	9
46	6628	6637	6646	6656	6665	6675	6684	6693	6702	6712	1	2	3	4	5	6	7	7	8
47	6721	6730	6739	6749	6758	6767	6776	6785	6794	6803	1	2	3	4	5	5	6	7	8
48	6812	6821	6830	6839	6848	6857	6866	6875	6884	6893	1	2	3	4	4	5	6	7	8
49	6902	6911	6920	6928	6937	6946	6955	6964	6972	6981	1	2	3	4	4	5	6	7	8
50	6990	6998	7007	7016	7024	7033	7042	7050	7059	7067	1	2	3	3	4	5	6	7	8
51	7076	7084	7093	7101	7110	7118	7126	7135	7143	7152	1	2	3	3	4	5	6	7	8
52	7160	7168	7177	7185	7193	7202	7210	7218	7226	7235	1	2	2	3	4	5	6	7	7
53	7243	7251	7259	7267	7275	7284	7292	7300	7308	7316	1	2	2	3	4	5	6	6	7
54	7324	7332	7340	7348	7356	7364	7372	7380	7388	7396	1	2	2	3	4	5	6	6	7

TABLE IV—*Continued*

LOGARITHMS 100 TO 1000

	0	1	2	3	4	5	6	7	8	9	1	2	3	4	5	6	7	8	9
55	7404	7412	7419	7427	7435	7443	7451	7459	7466	7474	1	2	2	3	4	5	5	6	7
56	7482	7490	7497	7505	7513	7520	7528	7536	7543	7551	1	2	2	3	4	5	5	6	7
57	7559	7566	7574	7582	7589	7597	7604	7612	7619	7627	1	2	2	3	4	5	5	6	7
58	7634	7642	7649	7657	7664	7672	7679	7686	7694	7701	1	1	2	3	4	4	5	6	7
59	7709	7716	7723	7731	7738	7745	7752	7760	7767	7774	1	1	2	3	4	4	5	6	7
60	7782	7789	7796	7803	7810	7818	7825	7832	7839	7846	1	1	2	3	4	4	5	6	6
61	7853	7860	7868	7875	7882	7889	7896	7903	7910	7917	1	1	2	3	4	4	5	6	6
62	7924	7931	7938	7945	7952	7959	7966	7973	7980	7987	1	1	2	3	3	4	5	6	6
63	7993	8000	8007	8014	8021	8028	8035	8041	8048	8055	1	1	2	3	3	4	5	5	6
64	8062	8069	8075	8082	8089	8096	8102	8109	8116	8122	1	1	2	3	3	4	5	5	6
65	8129	8136	8142	8149	8156	8162	8169	8176	8182	8189	1	1	2	3	3	4	5	5	6
66	8195	8202	8209	8215	8222	8228	8235	8241	8248	8254	1	1	2	3	3	4	5	5	6
67	8261	8267	8274	8280	8287	8293	8299	8306	8312	8319	1	1	2	3	3	4	5	5	6
68	8325	8331	8338	8344	8351	8357	8363	8370	8376	8382	1	1	2	3	3	4	4	5	6
69	8388	8395	8401	8407	8414	8420	8426	8432	8439	8445	1	1	2	2	3	4	4	5	6
70	8451	8457	8463	8470	8476	8482	8488	8494	8500	8506	1	1	2	2	3	4	4	5	6
71	8513	8519	8525	8531	8537	8543	8549	8555	8561	8567	1	1	2	2	3	4	4	5	5
72	8573	8579	8585	8591	8597	8603	8609	8615	8621	8627	1	1	2	2	3	4	4	5	5
73	8633	8639	8645	8651	8657	8663	8669	8675	8681	8686	1	1	2	2	3	4	4	5	5
74	8692	8698	8704	8710	8716	8722	8727	8733	8739	8745	1	1	2	2	3	4	4	5	5
75	8751	8756	8762	8768	8774	8779	8785	8791	8797	8802	1	1	2	2	3	3	4	5	5
76	8808	8814	8820	8825	8831	8837	8842	8848	8854	8859	1	1	2	2	3	3	4	5	5
77	8865	8871	8876	8882	8887	8893	8899	8904	8910	8915	1	1	2	2	3	3	4	4	5
78	8921	8927	8932	8938	8943	8949	8954	8960	8965	8971	1	1	2	2	3	3	4	4	5
79	8976	8982	8987	8993	8998	9004	9009	9015	9020	9025	1	1	2	2	3	3	4	4	5
80	9031	9036	9042	9047	9053	9058	9063	9069	9074	9079	1	1	2	2	3	3	4	4	5
81	9085	9090	9096	9101	9106	9112	9117	9122	9128	9133	1	1	2	2	3	3	4	4	5
82	9138	9143	9149	9154	9159	9165	9170	9175	9180	9186	1	1	2	2	3	3	4	4	5
83	9191	9196	9201	9206	9212	9217	9222	9227	9232	9238	1	1	2	2	3	3	4	4	5
84	9243	9248	9253	9258	9263	9269	9274	9279	9284	9289	1	1	2	2	3	3	4	4	5
85	9294	9299	9304	9309	9315	9320	9325	9330	9335	9340	1	1	2	2	3	3	4	4	5
86	9345	9350	9355	9360	9365	9370	9375	9380	9385	9390	1	1	2	2	3	3	4	4	5
87	9395	9400	9405	9410	9415	9420	9425	9430	9435	9440	0	1	1	2	2	3	3	4	4
88	9445	9450	9455	9460	9465	9469	9474	9479	9484	9489	0	1	1	2	2	3	3	4	4
89	9494	9499	9504	9509	9513	9518	9523	9528	9533	9538	0	1	1	2	2	3	3	4	4
90	9542	9547	9552	9557	9562	9566	9571	9576	9581	9586	0	1	1	2	2	3	3	4	4
91	9590	9595	9600	9605	9609	9614	9619	9624	9628	9633	0	1	1	2	2	3	3	4	4
92	9638	9643	9647	9652	9657	9661	9666	9671	9675	9680	0	1	1	2	2	3	3	4	4
93	9685	9689	9694	9699	9703	9708	9713	9717	9722	9727	0	1	1	2	2	3	3	4	4
94	9731	9736	9741	9745	9750	9754	9759	9763	9768	9773	0	1	1	2	2	3	3	4	4
95	9777	9782	9786	9791	9795	9800	9805	9809	9814	9818	0	1	1	2	2	3	3	4	4
96	9823	9827	9832	9836	9841	9845	9850	9854	9859	9863	0	1	1	2	2	3	3	4	4
97	9868	9872	9877	9881	9886	9890	9894	9899	9903	9908	0	1	1	2	2	3	3	4	4
98	9912	9917	9921	9926	9930	9934	9939	9943	9948	9952	0	1	1	2	2	3	3	4	4
99	9956	9961	9965	9969	9974	9978	9983	9987	9991	9996	0	1	1	2	2	3	3	3	4

TABLE V.—Compound Interest

Amount of $1, at compound interest for periods 1 to 50 at various *periodic rates.

Periods. *n*	*Periodic Rates.							
	2%	3%	3½%	4%	4½%	5%	6%	7%
1	1.02000	1.03000	1.03500	1.04000	1.04500	1.05000	1.06000	1.07000
2	1.04040	1.06090	1.07123	1.08160	1.09203	1.10250	1.12360	1.14490
3	1.06121	1.09273	1.10872	1.12486	1.14117	1.15763	1.19102	1.22504
4	1.08243	1.12551	1.14752	1.16986	1.19252	1.21551	1.26248	1.31080
5	1.10408	1.15927	1.18769	1.21665	1.24618	1.27628	1.33823	1.40255
6	1.12616	1.19405	1.22926	1.26532	1.30226	1.34010	1.41852	1.50073
7	1.14869	1.22987	1.27228	1.31593	1.36086	1.40710	1.50363	1.60578
8	1.17166	1.26677	1.31681	1.36857	1.42210	1.47746	1.59385	1.71819
9	1.19509	1.30477	1.36290	1.42331	1.48610	1.55133	1.68948	1.83846
10	1.21899	1.34392	1.41060	1.48024	1.55297	1.62889	1.79085	1.96715
11	1.24337	1.38423	1.45997	1.53945	1.62285	1.71034	1.89830	2.10485
12	1.26824	1.42576	1.51107	1.60103	1.69588	1.79586	2.01220	2.25219
13	1.29361	1.46853	1.56396	1.66507	1.77220	1.88565	2.13293	2.40985
14	1.31948	1.51259	1.61870	1.73168	1.85194	1.97993	2.26090	2.57853
15	1.34587	1.55797	1.67535	1.80094	1.93528	2.07893	2.39656	2.75903
16	1.37279	1.60471	1.73399	1.87298	2.02237	2.18287	2.54035	2.95216
17	1.40024	1.65285	1.79468	1.94790	2.11338	2.29202	2.69277	3.15882
18	1.42825	1.70243	1.85749	2.02582	2.20848	2.40662	2.85434	3.37993
19	1.45681	1.75351	1.92250	2.10685	2.30786	2.52695	3.02560	3.61653
20	1.48595	1.80611	1.98979	2.19112	2.41171	2.65330	3.20714	3.86968
21	1.51567	1.86029	2.05943	2.27876	2.52024	2.78596	3.39957	4.14057
22	1.54598	1.91610	2.13151	2.36991	2.63365	2.92523	3.60354	4.43041
23	1.57690	1.97358	2.20611	2.46471	2.75217	3.07152	3.81976	4.74054
24	1.60844	2.03279	2.28332	2.56330	2.87602	3.22510	4.04894	5.07237
25	1.64061	2.09378	2.36324	2.66583	3.00544	3.38635	4.29188	5.42744
26	1.67342	2.15659	2.44595	2.77246	3.14068	3.55567	4.54939	5.80736
27	1.70689	2.22129	2.53156	2.88336	3.28201	3.73346	4.82224	6.21388
28	1.74103	2.28792	2.62016	2.99870	3.42970	3.92013	5.11170	6.64885
29	1.77585	2.35656	2.71187	3.11864	3.58406	4.11614	5.41840	7.11427
30	1.81134	2.42726	2.80672	3.24339	3.74532	4.32194	5.74351	7.61227
31	1.84759	2.50008	2.90501	3.37312	3.91386	4.53804	6.08812	8.14513
32	1.88454	2.57508	3.00670	3.50805	4.08998	4.76494	6.45340	8.71529
33	1.92224	2.65233	3.11193	3.64837	4.27403	5.00319	6.84061	9.32536
34	1.96068	2.73190	3.22085	3.79430	4.46637	5.25335	7.25115	9.97813
35	1.99989	2.81386	3.33358	3.94608	4.66735	5.51600	7.68611	10.6766
36	2.03989	2.89827	3.45025	4.10392	4.87738	5.79182	8.14728	11.4240
37	2.08069	2.98518	3.57101	4.26806	5.09686	6.08141	8.63611	12.2236
38	2.12230	3.07478	3.69599	4.43880	5.32618	6.38548	9.15428	13.0793
39	2.16475	3.16702	3.82535	4.61635	5.56590	6.70475	9.70354	13.9948
40	2.20801	3.26203	3.95924	4.80100	5.81637	7.03999	10.2855	14.9745
41	2.25221	3.35989	4.09781	4.99306	6.07811	7.39199	10.9029	16.0227
42	2.29725	3.46069	4.24124	5.19276	6.35162	7.76159	11.5571	17.1443
43	2.34320	3.56451	4.38968	5.40047	6.63744	8.14967	12.2505	18.3444
44	2.39006	3.67144	4.54332	5.61649	6.93613	8.55715	12.9855	19.6285
45	2.43786	3.78159	4.70233	5.84115	7.24826	8.98504	13.7647	21.0025
46	2.48662	3.89503	4.86692	6.07480	7.57443	9.43426	14.5906	22.4727
47	2.53635	4.01188	5.03726	6.31779	7.91528	9.90597	15.4660	24.0458
48	2.58708	4.13224	5.21356	6.57050	8.27146	10.4013	16.3939	25.7290
49	2.63882	4.25621	5.39604	6.83330	8.64368	10.9213	17.3776	27.5300
50	2.69160	4.38389	5.58491	7.10665	9.03265	11.4674	18.4202	29.4571

* Periods may be annual, semi-annual or quarterly, etc. Periodic rates are proportioned to the length of the period. Thus, 4% annual = 2% semi-annual rate.

INDEX